How to Choose and Use

Lumber, Plywood, Panelboards and Laminates

by Mel Marshall

Drawings by
Vantage Art, Inc.

POPULAR SCIENCE

HARPER & ROW

New York, Evanston, San Francisco, London

Manufactured in the United States of America

Contents

ACKNOWLEDGMENTS

For individual help, spending time replying to queries, suggesting research sources, providing data, product information, drawings and photos, I'm very grateful to Pam Allsebrook, Rich Brill, Lenora Cerrato, S. T. Clark, Linda Doneff, Sharleen Escudero, Maryann Ezell, Sean Fitzgerald, Jim Gundy, Jim Henjum, Don Holcomb, Jim Kidd, Jim Nolan, Richard Silver, DeLos Snodgrass, James Steward, Sr., and Howard Williams.

My sincere thanks go also to the trade associations and firms which provided such a wealth of data and whose courtesies included permission to reproduce copyrighted material: American Hardboard Association, American Plywood Association, American Walnut Association, Appalachian Hardwood Manufacturers, Inc., California Redwood Association, National Association of Plastics Fabricators, National Forest Products Association, National Particleboard Association, Northeastern Lumber Manufacturers Association, Celotex Company, Formica Corporation, U. S. Gypsum Company, and Ralph Wilson Plastics Company.

Introduction

Newcomers to building and home craftwork know that there are specific materials best suited to any given job. Even veterans in these crafts know that with the increasing growth of products it's becoming harder and harder to make the best choice.

In many cases, of course, it's not really necessary to be overly choosy. If you're putting up a shelf inside a closet, it doesn't make a great deal of difference whether you use white pine, yellow pine, redwood, ash, oak, mahogany, plywood, or particleboard. Nor does it matter, if you choose lumber for such a job, exactly what grade you ask for, especially if the shelf is high and will be hidden by a door that's kept closed most of the time.

You will know the difference in your pocketbook. There's a wide variance in prices between grades of lumber, as well as in types of panelboards. There's no point in spending the extra cash for a #1 or #2 grade lumber for a seldom-seen shelf in a closet when #3 or #4 grade will do the job equally well.

Remember, though, that the reverse is also true. If, for example, you're building a storage room onto your garage or your house, you are starting the job at ground level. What you select for the timbers that form the addition's foundation, for the stringers that circle a concrete foundation, or for the joists that support the floors may make a great deal of difference.

You not only need to know whether the span you plan will need 2×12s to bear the load they'll be expected to carry, or whether you'll be safe in using 2×6s. You might find that in your area treated sill plates and floor joists are necessary because of humidity or soil conditions or termite infestation.

If you're putting in a room partition, you want studding that will be straight to begin with and will stay straight for the life of the partition—studs that won't develop a diagonal warp and cause cracks to form in your newly installed drywall. In this case, you not only should be concerned with the grade of the studding but should check the ends of the studs to find out whether they were flat-sawed or quarter-sawed. Or did you know that some flat-sawed woods tend to warp in a diagonal twist even if they've been very carefully seasoned? Do you know what flat-sawing and quarter-sawing mean?

What's true of any purchase you make is equally true of lumber: an informed buyer is the best guarantee of getting satisfactory merchandise.

In years past, when an uninformed customer went shopping for building material, he could be reasonably sure that he'd be served by someone better informed than he was about the product. Usually, the first

1

question that was asked of the unsure customer was: "Just exactly what do you plan to use this for?" The answer almost invariably brought suggestions from the salesman as to the products that would do the job best and at the lowest cost.

Unhappily, this is no longer the situation that meets the customer going into a lumberyard or building-supply center. Self-service is the watchword, just as it is in most other stores. The customer is free to browse and select what he wants, but eight out of ten of the salespeople know as little about the products they handle as the least-informed of their customers. Arming yourself with knowledge is simply a common-sense precaution that saves dollars and assures you of getting exactly what you need.

This book won't have all the answers, but it does have a lot of them. In it you just might find the reasons for some of the things you've been curious about but haven't gotten around to asking. If it doesn't have all the answers to your specific questions, it will very probably suggest a source where you can get the answers.

Good luck on your building and workshop projects!

1 | A Brief History of Lumbering

WHEN THE FIRST settlers arrived in North America, they found a land so rich in trees that they attached little value to them. In fact, the word "lumber" illustrates this. Then as now, in British usage the word is applied to things of small value, household discards and litter. In England a storage closet for articles of no immediate use is called a "lumber room." During the years of settlement, logs and rough-hewn timbers formed the ballast for ships returning empty to Europe. British sailors began calling their ballast "lumber," and gradually the word became the American description for all kinds of boards and wooden timbers. (If you're interested in going a bit further, "lumber" derives from "Lombard," a Teutonic tribe of the Middle Ages whose members were considered shiftless and worthless by early Britons.)

In actual fact, early settlers in North America were little accustomed to having boards and timbers with which to build. In Europe, the kings and nobles owned the forests and all the trees in them. Cutting was carefully limited to trees required for use in castle or mansion; foresters patrolled the woods and kept serfs and commoners from poaching by cutting trees just as gamekeepers kept them from killing the nobility's deer for food.

An echo of this is found as late as 1880, when Yeats wrote of his rude hut in Innisfree, "of clay and wattles made." Wattles are twigs, withes, which were woven into a sort of latticework and then plastered with mud to form the walls and roofs of the hovels in which the peasants lived. The wood from the annual harvest of trees was used to make furniture for the castle, hafts for spears and other weapons, carts and saddlery, and firewood. Part of the timber was sold in the cities, but the peasantry saw little of the forest's harvests in the days when North America was being settled.

On this side of the Atlantic, the newly arrived immigrants found timber that could be had for the cutting. The newcomers who pushed past the sparsely settled Atlantic coastal strip and moved into the wilds of New England's interior, to the forests of Pennsylvania, Tennessee and Kentucky found more trees than they really wanted. After clearing enough ground for a cabin, the average settler then chopped and burned the timber standing close to his new home; he needed a clear area all around it so that the Indians would have no cover from which to launch a surprise attack. Later, still more ground had to be cleared for crops.

About the only prohibition on cutting that was imposed on the British colonists was a ban of felling any trees marked with the king's brand. The tallest, straightest pines and the biggest oaks were chosen by timber

3

Appalachian Hardwood Manufacturers Assn.

In the wilds of New England's interior, and in the forests of Pennsylvania and Kentucky, settlers found deciduous trees such as oak, maple, and walnut in almost endless stands, as well as pines, cedars and spruce.

Later, as settlement moved West, great ranges of mountains covered by thick growths of conifers were found: fir, pine, redwood.

American Plywood Assn.

cruisers from the British navy, which needed wood for its growing fleet. Even when the trees harvested by shipbuilders for navy and merchant vessels had been cut, there were still great forests to be chopped by anybody who wanted to sell trees to the crude mills that began operating toward the end of the seventeenth century.

These mills had little resemblance to the wood-products factories of today. They were tiny operations. Most of them operated two or perhaps three pit saws, which were just what their name indicates. The saw was a vertical blade at one end of a pivoted beam which extended over a pit in which the sawyer and his helper worked. They pulled the saw down, a counterweight on the other end of the beam raised the blade, a work ox or mule dragged the log forward over the pit, and the log was reduced to squared timbers. These went to other pit saws to be reduced to planks.

Smaller versions of the pit saw, called whipsaws, were used in the larger mills to cut the wide planks—some of them 3 feet wide—into what today would be called yard lumber. The whipsaws also cut the planks into strips—shop lumber—which went to the cabinetmakers for use in furniture.

Pit saws, operated by muscle-power alone, were the tools of the earliest sawmills.

Even before these first mills sprang up in the New England states, even before the days of the pit saw, there were a few hardy timber harvesters who produced boards and timbers by hand. Using wide wedges and sledges, they split the logs they felled into boards, and then worked the faces of the boards smooth with axes and adzes. It took great skill and a disregard for toes to use an adze. These tools are still employed on occasion, but today their use is rare and confined to squaring and surfacing big timbers. Few timberworkers today have even seen an adze, which is a type of ax with a curved blade set at right angles to the handle. The adzeman stands on the timber or board and swings the adze between his feet in its chopping stroke.

Water-powered lumber mills were born in America, where both timber and streams were plentiful and found close to one another. New England, especially Maine, was the center of the infant lumber industry of colonial days, though an early mill was in operation in Virginia in the mid-1600s. Like all water mills, the early lumber mills had a huge wheel suspended in a millrace at one side of a dammed-up stream. Hand-cut wooden cogwheels at the waterwheel's hub transmitted the power generated as the wheel turned into belt- or cog-driven up-and-down saws in the millhouse on the bank. Circular saws appeared about 1830, and during this same decade the sawdust, bark, slabs, and similar mill waste began to be used to generate steam power for the operation of the mills. The gang saw, a line of as many as twenty-four circular saws, parallel-mounted on a common hub or axle, was in use by the 1860s, and in the 1880s the modern band saw came into general use. Gradually, the size of the mills increased and with it their hunger for logs.

Few timberworkers today have even seen an adze, which is swung by a man standing with his feet spread apart on the log, chopping toward himself. Adzemen usually ended their lives with fewer toes than they had when born.

LOGGING AND MILLING IN MODERN TIMES. Conservation in today's meaning of the word was never thought of in those beginning days, but logging practices of the era brought about an unconscious conservation. Because there were so many trees, the loggers ignored all but mature specimens. They cut these and moved on; often, the mill moved with them. The saplings and immature trees then were left to reach maturity. Clear-cutting, a later practice of the age of mechanized logging that began just before the turn of the century, was not commonplace in logging's early period.

During logging's early days, the industry was centered in the New England states, especially in Maine. By 1840, the center had shifted to New York State, where it stayed for about twenty years, when Pennsylvania became the most important lumber state. Michigan took the lead by the early 1860s, then surrendered it for a couple of years to Minnesota. Minnesota in turn was supplanted by Wisconsin, which remained the lumber center of the nation until the turn of the century.

In the beginning of the 1900s the western timberlands in Washington and Oregon were vying for lumbering leadership with Louisiana, where huge stands of yellow pine were being harvested. Then, in the late 1930s, Oregon became the nation's number-one lumber-producing state, and it remains the leader today. Throughout the period from the beginning of logging, however, the southeastern and mid-south areas consistently produced great quantities of oak as well as smaller amounts of other hard woods. They also yielded, as they still do today, a continuing flow of yellow pine to supplement the diminishing quantity of white pine that is produced in other logging areas of the nation.

Timber production from the days of the first colonists until the years following the Civil War, and even into the 1890s in the northern and western areas, changed very little. Logging was a matter of men's muscles, aided by horsepower, mulepower, and oxpower until the beginning of the twentieth century. The legends of Paul Bunyan and Babe, his big blue ox, had more than a core of fact in their forming, for mechanization moved into the woods with comparative slowness.

Hand axes and hand saws, wedges and mallets, were used to fell the trees. Skid roads—courses of saplings and branches laid on the ground on which tree trunks were dragged by the brute force of mules, oxen, or horses—were used to get the huge logs to a staging point, river or railroad, from where they were hauled to the mills. Lumbermen stayed in the timber in rough camps throughout the felling season, working a seven-day week, with holidays on Christmas and the Fourth of July.

MACHINES COME TO THE FOREST. Logging mechanization began with donkey engines, small compact steam engines with vertical boilers. Fired with wood slash, the boilers and engines were skidded from one timber stand to the next, much as logs were skidded from the stands to the nearest point where they could be easily transported to the mills. The donkey engines operated until true mechanization developed in the period after World War I, when the internal-combustion engine made possible powerful and agile machines light enough and maneuverable enough to be operated on rough terrain. The crawler, or continuous-tread tractor, began to predominate in logging work. This tractor was a direct relative of World War I's small experimental tanks; there is still a question whether the tank or the crawling-type tractor appeared first.

There were some steam tractors in the woods as early as the 1870s. These were the same type of machines used in agriculture, particularly in wheat farming in the Great Plains states, but they adapted badly to forest operation. On the smooth and level prairies, moving generally in a fairly straight line, the steam tractor worked well. In logging terrain, which is more often than not rough, wet, and hilly, the steam tractors proved to be too slow, too clumsy, and too heavy to give satisfactory service.

With the advent of the crawler tractor and the internal-combustion engine, with its high horsepower-to-weight ratio, the era of truly mechanized logging arrived. Today, the crawler tractor is supplemented by wheeled tractors and other equipment that moves across any terrain on giant low-pressure tires. Huge trucks haul logs direct from the forest to the mill; the trucks are loaded by claw lifts or cranes which are themselves descendants of the early machines. Logging is now less a matter of individual human or animal muscle than it is of skill with machines.

Huge trucks are loaded with giant logs by claw-lifts or cranes.

California Redwood Assn.

The process begins with the modern chain saw, which has for all practical purposes replaced the ax, felling saw, bucksaw, and other hand tools, and reaches all the way to the sawmill.

Mill mechanization antedated the arrival of machines capable of working in the forests. Increasingly heavy and complex milling machinery, such as ganged band saws which would reduce a huge log to boards or timbers in a single pass, giant planers, and similar equipment, ended the early practice of sawmills following the logging stands. This was practical—in fact, easy—only as long as a sawmill consisted of a few pit saws and whipsaws operated by gravity plus manpower. The first anchoring of sawmills came when waterpower began to replace manpower in the sawing, but even then the first large mills—large by the day's standards—continued to produce only rough-sawn, green lumber. The seasoning of the rough-cut wood and then its finishing into boards and other shapes was not carried out at early sawmills; they were *saw*mills in the strict sense of the word.

Even by the early 1800s the mills needed permanent sites to provide waterpower and to accommodate their batteries of saws, which were quickly joined by milling machines. Much lumber was milled when green in the industry's beginning days, and later seasoned and resawed or remilled to the desired dimensions. The work sequence was from sawmill to drying yard—later to drying kiln—and then to a finishing mill which did the necessary planing and shaping.

For all practical purposes, chainsaws and wedges have replaced felling-axes and bucksaws powered only by men's muscles.

California Redwood Assn.

Securely anchored now, rather than being transient like its primitive ancestors, the sawmill of today does far more than just convert logs into rough-cut planks and timbers. It is a woodworking factory, highly mechanized and automated, integrated with the procurement of logs from the timber stands. The mill of today is no longer just a sawmill. It produces finish lumber in quantities that would have been unimaginable to lumbermen of an earlier day.

Most of the large modern mills receive a continuous supply of logs which are rough-sawn and then moved to a drying yard or kiln. When ready for further working, the rough stock is returned to the mill and converted into boards, dimensional lumber, and timbers. In addition, most large mills turn out finish lumber, pattern stock such as bungalow and bevel siding, shiplap siding, tongue-and-groove lumber, and architectural trim and moldings.

Mills in locations where their timber supplies are of suitable types produce not only the kinds of lumber just mentioned, but also box stock, furniture frames, plywood cores, and veneer facings. Eastern, Southeastern, and mid-South mills, in the hardwood and yellow pine zones, also turn out special shapes that ultimately become tongue-and-groove flooring, barrel staves, tool handles, gunstocks, bowling pins, baseball bats, luggage, toys, and so on through a long and varied list.

However, the major output of almost all mills continues to be what the industry describes as construction lumber. This includes not only boards, which in industry terminology means square-edged lumber less than 1 inch thick, but dimension lumber, which is between 2 and 4 inches thick, and timbers, which are 5 inches or more in the smallest dimension. Even though most mills produce other products as well, the lumber used in construction remains their basic output.

Very rarely today do we see the old production cycle of a sawmill confined to the production of rough-sawn boards and timbers, which are then shipped elsewhere for drying and still somewhere else to be planed or shaped. Planing mills do still operate, but the trend of the industry is toward large, fully integrated mills which not only have a greater capacity in terms of total board feet of output, but also turn out a greater variety of products.

ANATOMY OF A TREE. Like most of the knowledge mankind has gained of natural things, that of today's forestry technicians was gained in the school of experience. The first step in learning was to acquire an understanding of how nature creates trees. The knowledge of a tree's anatomy gained by study and observation enables us to use each portion of wood derived from lumbering to the greatest advantage, not just in boards and timbers, but in lumber's by-products as well. The illustrated cross section showing the internal structure of a tree is worth studying, because it will, when interpreted, show why wood behaves as it does in use.

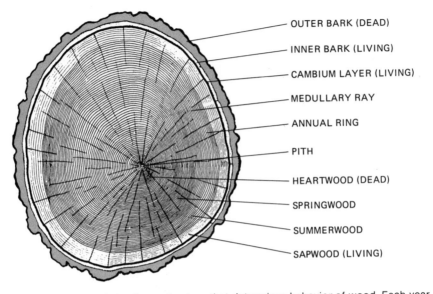

OUTER BARK (DEAD)

INNER BARK (LIVING)

CAMBIUM LAYER (LIVING)

MEDULLARY RAY

ANNUAL RING

PITH

HEARTWOOD (DEAD)

SPRINGWOOD

SUMMERWOOD

SAPWOOD (LIVING)

Cross section of tree shows inner structure that determines behavior of wood. Each year the inner cores of pith and heartwood increase in size as the tree's diameter grows. There is a difference between the characteristics of lumber cut from heartwood and that cut from the outer area, which is made up of summer and spring growth; cells in springwood are formed at a period when the tree is sucking up larger quantities of water and circulating more sap during its greatest growth period. The springwood cells are larger than those of summerwood. The heartwood is no longer growing in cell formation, but is valuable as lumber.

Anybody who's ever struggled to make a warped board lie flat knows the problems such boards present to the user. Not everyone using lumber realizes, though, that warping doesn't take place according to some random whim, but on a reasonably predictable pattern. The diagrams on page 12 of the cross section of a log illustrate these patterns. You will see that each of the six patterns is different. They would recur with the same differences if the log were turned 45° or 90° and the position of the diagrams remained the same in relation to the woodgrain, which is created by a tree's annual growth rings.

In diagrams A and A1, the warps follow the same curve, but a board cut from the section containing A can be expected to warp more than the board cut along the grain pattern A1. A board or timber cut from C will have a different grain pattern from one cut as is C1. The timber marked C will shrink uniformly when dried; that from C1 will tend to twist. A board cut to show the grain pattern B will shrink uniformly; one cut to have pattern B1 will crown in the center. A circle cut as is D will dry down to an oval shape.

Warping results from the cells of wood drying in different degrees when its moisture is removed in curing. The wood structure of the sap-filled growth rings is denser than that of the pithy pulp between them. Even the best-seasoned boards will warp to some extent, but the warp-

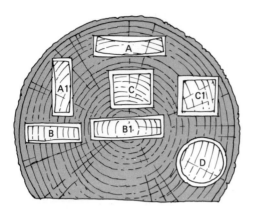

Here, superimposed on a cross-section of a tree trunk, are the warping characteristics of boards or timbers cut from different parts of the log. The white areas show the patterns of shrinkage. The text explains why these typical shrinkage patterns cause boards to warp in a predictable manner.

ing can be minimized by sawing the log in specific planes in relation to the grain. There are three types of sawing: flat, quarter, and rift. These terms describe in general lumber terms the angle of the grain in relation to the saw at the time a log is cut into boards.

In *flat-sawing*, the cuts are made in parallel, with the log in the same position. In *quarter-sawing*, the log is cut into quarters and each quarter is sawed separately, In *rift-sawing*, the saw cuts are made from triangular wedges of the log, with the blade at an angle of not less than 35° nor more than 65° to the annual rings.

Flat-sawing produces a greater number of boards from logs of the same relative size. As the diagrams showed, the few boards from the center of the log will be the most desirable. Actually, a flat-sawn log yields a percentage of boards equivalent to those produced by both quarter-sawing and rift-sawing. When a log is quartered or wedged to produce only rift-sawn or quarter-sawn boards, the log will yield fewer boards. Thus, flat-sawing is the most economical method to use.

Look now at the illustration showing how you can recognize boards that have been flat-sawn, quarter-sawn, and rift-sawn. A pattern characteristic of each shows in the board face and end grain. Flat-sawn boards will show the grain pattern in arcs on the end grain and long arrow-shaped lines on the faces. Quarter-sawn boards will show parallel grain marks on ends and faces. Rift-sawn boards will show quarter-arcs on ends and irregular grain lines on faces.

Finally, look at the warping patterns characteristic of each method of sawing. The grain pattern will help you select lumber that will warp the least and be far easier to use.

LUMBERING AND CONSERVATION. Lumbering and milling operations at one period of their history earned much criticism for being wasteful of resources. This period ended long ago, with the adoption of conservation measures that are now generally observed by all lumber

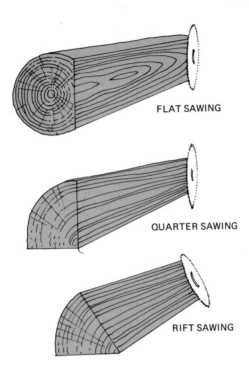

FLAT SAWING

QUARTER SAWING

RIFT SAWING

These are the relationships of log to saw blade in the three methods of sawing logs into boards. In flat-sawing, all cuts are parallel from one side of the log to the other. In quarter-sawing, the log is quartered and all cuts made parallel along each quarter of the log. In rift-sawing, the saw makes parallel cuts at an angle of not less than 35° nor more than 65°.

Flat-sawn boards will have a vee-shaped grain pattern on their faces and an arc pattern on their ends; the smallest arcs will begin on the heart side, the arc rising to the bark side. The heart side should always be nailed to the inner surface of construction. Quarter-sawn boards will have a parallel grain pattern on their faces and an end pattern at right angles to the faces. Rift-sawn boards show a wide-spaced parallel grain pattern on their faces and a slanting end-grain pattern.

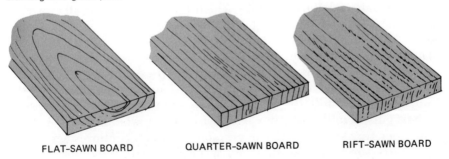

FLAT-SAWN BOARD   QUARTER-SAWN BOARD   RIFT–SAWN BOARD

operators. These measures begin in the forest and are carried through the entire production chain. As mentioned earlier, the first lumbermen practiced a conservation that was part expediency, part necessity. Given the conditions of the time in which they operated—lack of heavy-duty transportation and equipment capable of handling huge logs efficiently—early loggers cut only mature trees with a maximum size limit dictated by what their mules or oxen could drag and the early sawmills could handle. They left standing both the real forest giants and the seedlings and saplings, though inevitably a certain amount of smaller growth was destroyed when trees were felled and dragged into position for the haul to the mills. After harvesting the mature trees, the loggers moved on to a new stand and repeated the process; the sawmill usually moved with them. In time, the forest not only overcame the loss of its mature trees, but actually benefited by their removal.

What these early loggers worked in was a mature forest. This is the term applied to a growth of trees of pretty much the same size, all of them towering high enough to shut off the sunlight needed by seedlings to become saplings and for saplings to mature into trees. A mature forest will grow no more, no matter how long it is left untouched, as new growth becomes possible only when the mature trees die and fall. Then and only then can new trees survive in the area vacated by the fallen tree.

Lumbering removed only the mature trees and encouraged the forest to fresh growth. In time, the seedlings left by the loggers became saplings and the saplings grew into trees. This kind of growth is very closely akin to that dictated by nature, and it will not occur if the forest is destroyed by fire or is clear-cut in a logging practice that is still under debate. When an area has been clear-cut, or fire has swept through it, then man must step in to speed up nature's slow recycling process.

Today, in forest areas where only mature trees are to be harvested, foresters mark each tree that the loggers are to cut. This is in line with early practice, though today the reason is the knowledge gained from study and experience, not merely expediency or necessity. At the same time, in areas that have been clear-cut—which simply means that all growth is removed by logging—or devastated by fire, seedlings are being planted. And in areas where trees have never grown, tree farms are being planted. Thus, while the demand for lumber and wood products keeps increasing, the sources of these products keep expanding.

Although early sawmills came in for their share of accusations of wastefulness, modern mills have never really earned the blame. The very earliest pit-saw mills did. They left slash and debris behind them in their brief operating span, which ended more than two centuries ago. Their successors, the steam-powered mills, left no waste; they burned bark, slabs, and sawdust in their boilers. By the time waterpower replaced steam power, paper mills had come to most logging areas, providing a market for mill wastes. In later years the lumber industry has provided its own market for its waste, by processing sawdust, slabs, bark, and chips into various types of composition boards used by builders and home craftsmen. These include plywoods, particleboards, hardboards, and softboards, as well as such nonlumber products as gypsum board.

Foresters mark each tree that is to be harvested.

Trees are spot-felled to drop along a line that will do least damage to young growth.

Considering the size of some of the trees, especially in the Western redwood forests, line-felling requires nice judgment on the part of the felling crews.

California Redwood Assn.

Clear-cut areas, such as the one on the slope at right, and fire-devastated areas, are replanted with seedlings.

American Plywood

TREATED WOOD. Right now, though, let's stay with the mainstream of lumbering for a quick look at one specialized type of lumber which could be very important to you if you're planning to put up any type of building on soil that is swampy, in a climate that is unusually humid, or in an area that is termite-ridden, or if the building will be used for purposes that will expose its basic materials to constant or intermittent wetting. This is treated wood.

Long years ago, the builders of wharves and those who used wood in swampy or unduly wet areas discovered that pilings and foundations needed special protection. For a century or more, the only protection available for wood that came into intermittent contact with water, such as the pilings of wharves and piers, the ties used in laying railroad tracks, and the foundation pilings of buildings erected in swamplands, was to coat the wood with a coal-tar product such as asphaltum, or a wood-tar product such as creosote. These treatments gave little more than surface protection, however, and once the surface coating in a relatively small area was breached, the unprotected wood exposed began to deteriorate.

The next step forward was placing the wood requiring protection in a sealed chamber and driving the protective substances deeply into its grain by the use of heat and pressure. Still, wood treated by these methods had undesirable characteristics. The asphaltum or creosote compounds used emitted an unpleasant odor, made the wood difficult to work, and also soiled anything with which the treated wood came in contact.

In the early 1900s, new products to inhibit wood decay came from chemical laboratories in the form of metallic salts that gave better protection from attack on the wood's fibers and had no odor or stain-transferring qualities. Today, there is available a wide selection not only of treated timbers for piling and foundations, but of treated wood to be used as siding, flooring, joists, and beams of buildings and other structures. This wood resists the attacks of wet or acid soils and of termites and other insects. It is widely sold in the areas where such protection is required. You should certainly investigate treated wood if you plan to use wood in such a way that it will be exposed to intermittent or constant problem conditions.

PLYWOOD. Now, let's move on to building boards, as distinguished from boards sawed from boards sawed from trees. These are man-modified forms, of which plywood is perhaps the best known.

Plywood is an offshoot of the ancient art of veneering, which dates back more than three thousand years; veneered furniture has been found in the tombs of the pharaohs of Egypt and the emperors of China. It was a common art in Italy and France of the Renaissance, and in the great days of British cabinetmaking. Veneering is simply using a rare wood sliced into thin layers on the outside of a piece of furniture to cover a more common wood. Its sister craft is inlaying—creating a pattern or design of thin sheets of wood glued on a chest or table or wall panel. Both these crafts, like the making of plywood, involve the use of thin sheets of wood.

Certainly the idea of forming a uniform wood panel that would be free from warping and would be larger than any tree trunk could yield must have been tried unrecorded in the distant past. There are many examples of wood lamination, even among primitive tribes. The North American Indians laminated wood and horn to make their bows, and relics of other cultures show similar work.

In late-Victorian times, furniture makers in search of new ideas tried gluing thin wooden sheets around a metal form to make chair seats. These pieces are rare today, because the glues that were available then lacked permanence. The plies soon separated, and the furniture fell apart.

In a semiautomated plywood mill, panels travel through the edger to be stacked for shipping.

American Plywood Assn.

Modern plywood was first shown at the Lewis & Clark International Exposition in 1905. Panels in the exposition display were hand-crafted of Douglas fir by the Portland Manufacturing Co. of St. Johns, Oregon. The display sparked a flood of orders from manufacturers of doors, luggage, and furniture—and within a few years, automobiles. During the early days of the plywood industry, automobile running boards were a major market. In those first years, plywood manufacture centered in the area where the Douglas fir was a major timber tree—the Pacific Northwest. It is still the wood most widely used in plywood manufacture. The first successful plywood manufacturing process involved kiln-drying thin sheets of wood peeled from the fir trunks, then laminating them in a cross-grain pattern to a somewhat thicker core. The sheets of wood were called "plies," and hence the panels produced got the name "plywood."

Not all the earliest plywoods were an unqualified success. The process was still too new. Some of the glue mixtures bonded imperfectly; some of the plies were not dried enough and buckled. However, the manufacturing processes were refined and improved with experience, and by the early 1920s plywoods for interior use were giving satisfactory service.

In the early 1930s, the development of phenolic-resin glues made possible the fabrication of exterior-grade plywood. Using the new adhesives, it became possible to use plywoods in weather-exposed applications, and even in marine applications such as boat hulls. The famous PT boats of World War II had hulls fabricated from marine-grade plywood, and so did Howard Hughes' ill-fated flying boat, which got the nickname "Spruce Goose" from the plywood of which its hull and fuselage were made.

New glues, new methods of handling the plies, new presses used in the laminating, and, above all, the experience gained over decades have resulted in a range of plywoods of amazing versatility. There is quite literally a plywood for almost every building use, from water-resistant exterior sheathing to decorative interior paneling of exotic woods.

PANELBOARDS. Cousins to the plywoods, at least in form, are the panelboards made from forest-derived materials. These products vary widely in composition and in methods of fabrication. Like plywood, they are made up into panels of a standard size, usually 4×8 feet, and all of them owe their existence to wood in one form or another.

Those most limited in end use are the cellulose/asphaltic sidings. These are made from coarse, loosely laid sheets that are a cross between paper and cardboard, impregnated with an asphaltic binder and bonded under pressure into panels, usually 4×8 feet and of various thicknesses. They are used primarily for exterior sheathing, as they provide both a vapor barrier and a moderate degree of insulation. These panels also find limited use as subflooring over concrete; their resiliency underfoot makes a tile or roll vinyl floor covering much less tiring to walk on.

The next branch of the panelboard's family tree sprouts a number of softboards. These are basically a form of unusually thick cardboard —wood pulp dried and compressed rather than being produced in flexible sheets on calendering rolls as are most paper products. These softboards take a number of forms, but their relatively soft surface limits their application. In 4×8-foot and 2×4-foot panels, they may be used for above-wainscoting walls, but their most common use is in the form of ceiling acoustical tile. The softness of this material makes perforating or embossing very easy, and both panels and tiles come in many surface textures and designs.

Drywall paneling, also called gypsum board, is made by sandwiching gypsum sheets between layers of a very dense type of cardboard. By itself, gypsum is quite brittle; the outer cardboard layers both reinforce the gypsum core to prevent it from cracking and give it rigidity. The cardboard also makes it possible to nail the panels into place. There is a type of gypsum board for exterior use (see Chapter 5), but drywall paneling is used only as an interior wall sheathing. It can be finished by painting or plastering, or covered with wallpaper or fabric.

Much more versatile members of the composition board family are its next two members, particleboard and the hardboards. Particleboard is just what its name suggests: large panels, usually 4×8 feet, made from wood-waste particles: sawdust, chips, and splinters, mixed with a doughlike woodpaste and compressed under moderate heat into rigid sheets. The density or specific gravity of particleboard is greater than that of plywood, and this material also weighs more. It is available in various thicknesses and is very widely used not only in buildings, but in such applications as furniture cores with a veneer overlay, freestanding cabinets, and shelving.

Particleboard cures under pressure in one of the plants producing this wood composition board.

California Redwood Assn.

The hardboards are made from converted wood fibers bonded into panels under heat and pressure, with a lignin binding. The wood fibers are produced by whirling knives in what the manufacturers call an "attrition mill," or by loading wood chips into a high-pressure cylinder into which steam is fed gradually, then suddenly released. The "explosion" that follows the quick release reduces the steam-wet wood chips to a pulpy mass of fibers. These are then felted or matted and bonded by either a wet or a dry process.

In the wet process, water flowing through a pulp-laden mesh or screen aligns the fibers for their final felting, and the finished panel bears the mark of this screen in finished form. In the dry felting process, the fibers are aligned with airstreams. The final step in both these processes is to apply heat and pressure to weld the fibers together into panels. They are then placed in a conditioning chamber, which dries them.

At this stage, hardboard is ready for use in such applications as sheathing and underlay and for use by the home craftsman in various projects, but panels designed for interior finishing undergo further treatment. They may be textured or embossed in woodgrain or other designs; they may be coated with an impervious acrylic finish for use in finishing a kitchen or bath; they may be printed with a woodgrain or other designs; or a vinyl coating that combines finish and pattern may be laminated to one surface of the panel. Hardboard panels run neck and neck with plywoods as being the most versatile of all materials used today in construction and in home craftwork.

LAMINATES. Finally, we come to a nonlumber type of paneling: the laminates. These are semirigid plastic sheets used in finishing kitchen work surfaces, bathroom walls and cabinets, and furniture. Laminates have an interesting history. They started about as far from buildings as it's possible to get, having been developed in the early 1900s as a lightweight, easily formed replacement for heavy ceramic and mineral electrical insulators. Later, the material found additional uses for gears in autos and appliances and for coil cores and other parts of radios.

Laminates are made by bonding heavy paper cores between sheets of plastic resin, using heat and pressure. Decorative plastic laminates appeared when someone got the idea of printing a decorative design—instead of a part number or maker's trademark—on the paper core. The first widespread use of these new design-imprinted laminates was in the front panels of radio cabinets, and the next major use was that of marble-patterned laminate sheeting bonded to plywood to replace the heavy, scratch-prone marble traditionally used in soda fountains.

These were followed quickly by solid-color, fabric-patterned, and woodgrained laminates. The tough resin which forms the outside layer resists acids, alkalis, and scratches, and this makes laminates very popular as a surfacing material for such high-wear areas as kitchens and

Kraft paper, the "core" of laminates, passes through the phenolic resin treater bath as the first stop in the process that produces these versatile surfacing panels.

bathrooms. There is also a burn-resistant laminate, made by adding a layer of aluminum foil under the printed sheet that gives the material its surface design. The heat-conducting qualities of the aluminum prevent a dropped cigarette from burning a hole in the resin.

These are the materials you'll meet in later chapters. As we examine each of them, we'll be going into the details of their selection and the ways you can judge which is most suitable for whatever job you have on hand. We'll also cover the recommended methods of working them and applying them to your own projects.

GRADES, SPECIFICATIONS, AND STANDARDS. In virtually all the chapters that follow, you'll find a lot of figures mixed with descriptions. These give industry standards, establish differences in quality, or describe performance characteristics of various products and materials. Please don't pass these over lightly as being boring. They may be, but they're also guideposts to help you in choosing from among the sorts of materials to which they refer. More than that, if used properly they can indicate the selection of a less costly grade of lumber, or a lower-priced type of some processed products, that can be used in place of the higher-priced material you'd planned to use.

Being of a rather suspicious nature myself, I questioned at one time the validity of claims made by manufacturers or processors regarding the products they sell. These claims seemed to be self-serving by their very nature.

Experience and time have caused my skepticism to evaporate. The hard cold facts of life in a competitive economy are that no industry can afford to have its products get a bad name. An unfavorable reputation, justified or not, earned by one manufacturer within an industry ultimately is applied to all segments of that industry. There are no monopolies in the several industries represented by the producers and manufacturers of the products described in this book. Each manufacturer or producer is competing for his share of the market, and a decline in sales of the industry's products, caused by the bad reputation acquired by one firm within the industry, becomes a matter of industry-wide concern.

As a result, the industries which make up the field of lumber and building materials have chosen to police themselves. This works to the ultimate benefit not only of the industry, but of those who use its products. The specifications and grades and standards quoted in the pages that follow apply nationally. Often they contain recommendations that a specific type of product not be used in a certain manner, or for a certain application. This keeps the user from making mistakes which might be blamed on the processor or producer.

Many of the specifications are primarily intended to guide commercial builders in selecting products that will pass the many safety and structural-strength codes that are in force almost everywhere. By studying them and learning to use them, you will be able to choose not only the best products at the least cost for the job you have in mind, but you'll be secure in the knowledge that your job will be done right from the moment you begin to plan it.

2 | Choosing and Using Lumber

PERHAPS THE WAY to start is by getting acquainted with the special vocabulary used in the lumber industry. It will be used in describing sawing and milling operations and grading standards, and will be helpful to you when we get to talking about dimensions and methods of application.

At the beginning of the lumber-production chain, *loggers* in the forest *fell* trees with saws and axes and wedges. They clear the felled trunks of branches, then *buck* the trunk into *logs* of lengths that can be conveniently transported to a *sawmill*. At the mill, the logs are unloaded; if they are to be converted into lumber at once, the unloading is done at a *live deck*, from which the logs are moved to the mill. If the live deck is full, they go to a *cold deck* to wait until they can be handled. Just before going to the mill, the logs are *barked*; the bark is removed by high-pressure jets of water, by being grated against huge revolving drums, or by air hammers.

A *headrig* conveys the logs to the first saw, where a *sawyer* operating a band saw converts them into large rectangular timbers called *cants*. The cants are inspected in a preliminary grading, then a series of multiple band saws and edgers turns them into *rough-sawn green* boards. These pass on a *green chain* to inspectors who do another grading. Each board is pulled from the green chain and stacked according to grade.

After grading the boards move to one of two seasoning processes. Some may be *air-dried*, which means that each piece is stacked between spacers to let the moisture in it evaporate over a long period of time simply by exposing it to the air. Or, the green wood may be passed through a *dry kiln*, in which streams of heated air remove the moisture from the tree sap in a much shorter time. During the drying period, whether in a kiln or air drying, the wood shrinks appreciably.

Milling is the next step. After they are cool, the *seasoned* pieces go through *planers*, which smooth their surfaces and edges and reduce them to precise dimensions. Some of the planed pieces will go on to *shapers*, which convert them into *pattern lumber*, the name given to tongue-and-groove, shiplap, and other formed construction lumber. We'll meet these later on.

Others will emerge from the mill as *timbers, dimension lumber,* or *boards.* A second grading and perhaps a bit of additional milling will convert a limited quantity of the best boards into *finish lumber*, which is what you encounter when you go to your retail dealer. The final grading determines not only what you'll ask for to do the job you're working on, but the price you'll pay.

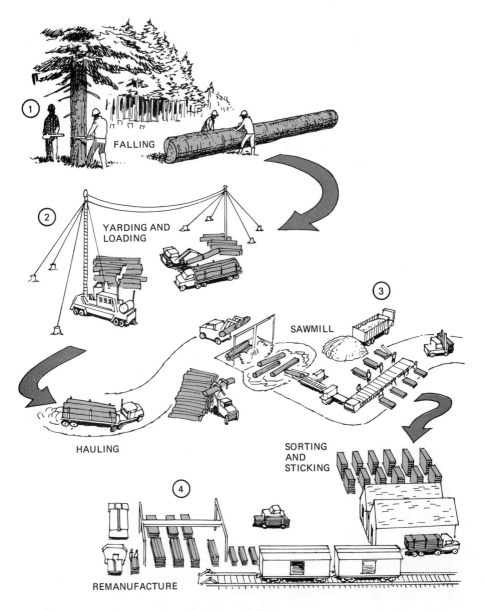

The steps in transforming trees into lumber products.

Now, let's backtrack a bit. At each working stage a bit of wood waste is produced: bark, sawdust, chips, odd pieces. This goes to mills or factories where it will be converted to other wood products: particleboard, composition board, paper.

Still backtracking, let's take a second look at the sawing step, to emphasize its importance. If you'll recall, there are three types of saw-

California Redwood Assn.

Logs in a millpond await debarking. The men standing at right in front of the truck give you an idea of the size of the logs being unloaded.

Revolving under the blade of a giant scraper, a log is stripped of its bark before going to the mill saws.

American Plywood Assn.

California Redwood Assn.

Dwarfed by the headrig on which he and the log ride, the sawyer turns a log being sawed into *cants,* big square timbers; these will later be sawed into boards.

Boards stacked for air-drying are covered and separated by spacers to give the air access to each of them. These stacks will stand for a year or more before being given a final grading and shipped.

California Redwood Assn.

ing, *flat*, *quarter*, and *rift*. You'll also recall that each method produces a different end-grain and face-grain pattern, as well as a different warping pattern: *cupping*, *twisting*, and *bowing* or *crooking*.

From a practical standpoint, cupping or twisting may in time cause subflooring to become uneven or studding to crack drywall partitions, and excessive crooking can throw courses of sheathing or siding out of line. These natural tendencies can be minimized by the proper application of lumber to the job, by proper nailing, and—most important—choosing the kind and grade of lumber the job calls for. All these will be covered later in this chapter.

BASIC DIFFERENCES IN LUMBER. Quite understandably, many people just getting acquainted with lumber become confused at once when they learn that woods classified as "softwoods" aren't always soft, and woods classified as "hardwoods" aren't always hard. Basswood is a good example. Its working characteristics are more like those of a good soft pine than any other wood, yet basswood is classified as a hardwood. This dichotomy extends to the pines. Southern yellow pine can be almost as hard as straight-grained maple, yet both hard and soft maples are included in the hardwoods, while yellow pine falls into the softwoods group.

Actually, the explanation is quite simple. Lumber is classed as hard or soft in reference to the type of foliage borne by the tree from which it comes. Deciduous trees—that is, trees which have flat spreading leaves—generally produce hardwoods. Coniferous trees—those which bear needles rather than leaves and whose seeds form around a core into a cone—go into the softwoods classification.

We need to face facts, though. When you go to buy lumber from a lumberyard or building-materials outlet, you're not likely to have a choice of more than three or four kinds. Lumber cut from trees that grow in countries other than the United States or Canada generally fall into the category of "exotic woods," and are for the most part sold only by specialty suppliers. There are good reasons for this. First, few of these woods are the types you'd be apt to ask for unless you were making furniture. For the average retailer, the volume of sales of exotic woods simply doesn't justify stocking them. There's another reason. The demand for exotic woods has always exceeded the supply, and even if a busy commercial lumber dealer wanted to carry them in stock, he'd be hard pressed to maintain a representative selection.

What you will find in the average lumber retailer's stock depends to a certain extent on where you live. In the East and Southeast, you'll very likely have a good selection of yellow pine, fir, and cypress. In the West and Southwest, you'll find white pine, fir, cedar, and redwood. Redwood is very widely distributed, and so is oak flooring. Most retailers try to keep a few pieces of walnut on hand, and in the East you'll have a wider choice of oak than just flooring. Of course, this isn't any sort of definitive description of lumber stocks. No two retailers go by quite the same guidelines.

There are good reasons for the average selection at retail outlets. Homebuilding and commercial construction are now large-scale businesses in most parts of the nation, and many builders buy direct from the mills, or from wholesale distribution centers. Retailers are left with the home-repair specialist and the small independent contractor who builds a house at a time instead of a hundred houses on a mass-production basis.

Given the changed pattern of construction that has come into being since World War II, the day of large and varied lumber stocks in average-sized retail outlets has passed. Local retailers are simply exercising good business judgment when they give priority to the merchandise they can move in profitable quantities, even though their return per board foot of hardwood is certainly greater than that per board foot of construction lumber.

TYPES OF LUMBER. To get back to the subject of softwoods and hardwoods, let's list them by name in each category. As the softwoods are more numerous, we'll look at them first, and in alphabetical order rather than by importance or predominance. We will also use their commercial names rather than identifying each of the species in each type. (Species information will be found in Appendix A, however.)

Softwood lumbers are cedar, cypress, fir, hemlock, juniper, larch, pine, redwood, spruce, tamarack, and yew.

Hardwoods are basswood, birch, maple (both hard and soft), oak (both red and white), poplar, and walnut.

If you think this is a limited list which omits a lot of kinds of wood that are pleasant to work with, you're right. And, at the risk of being repetitive, there's a reason. The woods just listed are those which are available in commercial quantities—quantities large enough to provide a steady supply to the major users of lumber.

There are indeed many other woods which are both beautiful and quite pleasant to work with. Cherry is one of them. Ash is another. Black walnut is still another. The list of woods that don't fall into the regularly available category could easily be extended to include many other familiar names. They would be kinds that are used in furniture and gunstocks, as veneer facing for paneling, and as inlays. And the chief reason why these kinds aren't on the list is that they are in short supply, and the limited amounts available are sold years before the trees are even harvested.

Now that we've identified the two kinds of lumber woods, let's go a step further and identify the jobs each of them does best. As we do this, you'll find that some woods can be used in many applications in construction as well as in home crafts. In the lists that follow, the woods are in alphabetical order, and the jobs or applications for each are in the order of greatest importance. It might be well to keep in mind that these lists sum up the judgment of the men who use wood in their daily occupations as well as those who cut and mill the logs from which lumber is made.

SOFTWOODS

Cedar

Closet linings
Roofing shingles and shakes
Veneer facings

Cypress

Fencing
Foundation posts
Railway ties
Interior wall paneling

Fir

Construction lumber (joists, rafters,
   stringers, siding)
Plywood
Laminated beams
Utility lumber, in grades #3 and #4

Hemlock

Plywood cores
Boxes and crates

Juniper

Fence posts
Telegraph poles
Interior finish moldings
Pencils

Larch

Construction lumber (joists, beams,
   stringers, siding)
Plywood
Telephone poles
Railway ties
Mine timbers

Pine, white

Construction lumber (rafters,
stringers, sheathing, siding, interior
trim molding, laminated beams,
subflooring, paneling)
Furniture frames
Boxes and crates, in grades #4 and
#5

Pine, yellow

Construction lumber (joists, rafters,
studding, rafters, sheathing, subflooring and flooring, structural
timbers)
Furniture frames
Shipping pallets, in grades #2, #3,
and #4
Boxes and crates, in grades #3 and
#4

Redwood

Construction lumber (exterior siding,
paneling, foundation posts)
Tanks
Shakes and shingles
Greenhouse benches and flooring
Pool enclosures and cabanas
Outdoor furniture

Spruce

Construction lumber (sheathing, sid-
ing, subflooring, interior and exteri-
or trim, architectural trim)
Boats and oars
Boxes and crates
Pulpwood

Tamarack

Construction lumber (joists, rafters,
stringers, siding)
Railway ties

Yew

Implement handles and wooden im-
plements
Cabinetry
Sports equipment, especially bows

HARDWOODS

Basswood

Interior trim moldings
Plywood cores
Furniture frames
Boxes and crates, in grades #3 and
    #4
Utility cabinets
Luggage frames

Birch

Architectural woodwork and interior
    trim moldings
Cabinetry
Paneling
Furniture

Maple, hard

Furniture
Paneling and interior trim
Cabinetry
Industrial flooring (ballrooms, gymna-
    siums, factories)
Implements

Maple, soft

Cabinetry
Toys
Interior paneling
Home and industrial flooring

Oak, red

Flooring and subflooring
Furniture
Tight barrels, for liquids
Implement handles
Paneling and interior trim

Oak, white

Flooring and subflooring
Furniture
Tight barrels, for liquids
Boats and ships
Paneling and interior trim
Implement handles
Cabinetry

Poplar

Weatherboard and siding
Furniture
Furniture frames
Interior trim
Paneling
Cabinetry

Walnut

Furniture
Cabinetry
Interior trim and paneling
Architectural woodwork
Gunstocks

Trees on the foregoing lists by no means exhaust the number of varieties that produce lumber or timber in small quantities or for special purposes. Some of their woods never reach the general lumber market, but are contracted far in advance of their annual cutting by furniture factories, veneer mills, sporting-goods makers, toymakers, toolmakers, and the large specialty houses that feature hardwoods for use by hobbyists and home craftsmen.

Cherry, for example, is highly prized by gunstock makers. It is also used where an exceedingly dense, dimensionally stable wood is required. It is the only wood used by the printing industry as a backing for picture plates. It is also used by toolmakers in the production of spirit levels.

Hickory goes into sporting goods, where it is valued for its great longitudinal strength. It is used in skis, gymnastic parallel bars, and other similar equipment. Hickory is also valuable as a furniture wood, and toolmakers use it widely in handles for hammers, garden tools, and similar items.

Butternut and sycamore are furniture and veneering woods, and butternut is also used in furniture and interior trim strips. Sycamore is used in the manufacture of slack cooperage (barrels in which dry materials are shipped in bulk).

CHOOSING THE RIGHT TYPE OF LUMBER. Even a casual look at the use lists will show that the uses to which a given kind of wood is put give a pretty good indication of the wood's characteristics.

Obviously, a wood that is recommended for foundation posts or telephone poles or railway ties will have the ability to resist the attacks of moisture, acid soil conditions, and insects and will be generally long-lived. Most such woods—cypress, juniper, larch, and so on—are seldom suitable for use as construction lumber. A notable exception to this, of course, is redwood, which is perhaps the most versatile of all softwoods in the applications for which it is suited.

Equally obviously, a wood such as walnut, with its density and fine graining, isn't a wood you'd use for such construction applications as sheathing or siding. Even though it would be quite satisfactory for such uses, its value per board foot is so much greater for specialty uses that using it as construction lumber would be profligate.

A look at the list of major applications for which yellow pine is best suited will show that it must have greater shear strength and grow to greater size than a wood like poplar. The yellow pine list indicates that it is widely used for beams, which indicates exceptional shear strength, and is also used for flooring which indicates surface toughness. Poplar, on the other hand, is best suited for applications requiring no great shear strength or surface hardness.

As still another example, any wood listed as being suitable for paneling must have an attractive, or at least an inoffensive, grain pattern. Such woods can be used either as solid panels or facings on plywood paneling, and you can logically expect to find them available in both solid and veneered forms.

Woods indicated as being widely used for fencing obviously have excellent weather-resisting abilities. So do those which are used for exterior siding. On the other hand, woods that are milled into interior trim or paneling won't always have the weathering qualities that would make them a good choice for outdoor use.

While the list on the preceding pages will give you most of the clues you need to make a wise selection of wood for both building and home craft projects, it's by no means complete. To provide an exhaustive compendium of all the woods and the applications for which they're suited would require a book the size of an unabridged dictionary. And, in order to use the list in choosing lumber, you need a bit more information about reading grade markings. This subject was passed over lightly earlier in the chapter, so let's take a closer look at it right now.

GRADING LUMBER. When you set out to buy a certain kind of lumber, armed with the information that's guided you to choose redwood instead of spruce or fir rather than hemlock, you still have one more factor to consider. That's the *grade* of lumber you're going to buy. Remember, the cost of two different boards or timbers of the same size and from the same kind of wood will depend on the grade given it at the mill.

More lumber is branded with a grade mark today than ever before, and this is a big advantage to you as a buyer. Of course, you need to be able to read the grade mark. The chances are it will be pretty much like the one illustrated, which identifies the source which established grade standards, the name or trade-association number of the mill from which the board came, the actual grade itself, the species or kind of wood from which it was sawed, and the moisture content,

Grading of lumber involves more than a hundred different factors. These take in such things as natural defects resulting from conditions under which the tree grew, defects that are the result of sawing or milling, variations in quality, and characteristics that the board bears that have resulted from a combination of natural and man-made causes. In grading a board for natural defects, there are eleven separate conditions of decay, sixteen for knots, twelve for pitch pockets. There are twenty manufacturing defects on the list of conditions the grader must note.

Among the natural defects are bark pockets, caused when a wound, perhaps a burn from a forest fire when the tree was young, resulted in a folding of the outer bark into the interior of the trunk. Pitch pockets occur naturally, when a deposit of sap is trapped between the tree's growth rings. Decay is usually caused by wood-destroying fungi; it is also called *dote* and *rot*. Some types of knots and their condition in the sawn board will create weak spots, while others do not. Most of these natural factors—those mentioned as well as others which are shown in the accompanying picture—enter into the grade determination.

So do man-caused blemishes, which are the result of sawing or milling. These come under the general heading of *mismanufacture,* and include scars or burns caused by the planer, skips and gouges that create a rippling effect on a board's surface, torn grain which occurs during finishing, areas of bark remaining on the board, and sawing that somehow produced boards uneven in width or thickness.

Typical of the grade marks branded on lumber is this one of the Northeastern Lumber Manufacturers Association. Here are what the markings mean:

S-Dry. The board has a maximum moisture content of 19% or less.
S-GRN. The board has a moisture content over 10%.
001. The producing mill's NELMA-assigned number, firm or brand name.
FINISH. The grade assigned to the respective board.
EASTERN WHITE PINE. The species of the respective board.

Northeastern Lumber Manufacturers Assn.

Here are the defects lumber graders look for during their inspections, and the causes of the defects. The more defects a board has, the lower its grade and selling-price. Unless the face of a board is to be exposed, as in a decorative interior use, the lower grades of lumber are quite satisfactory from the standpoint of strength and durability.

Northeastern Lumber Manufacturers

KNOTS: The most familiar natural characteristic, caused by sawing through a limb that grows from the heart of the tree.

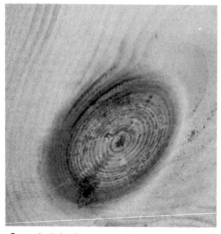
Sound, tight (intergrown), oval red knot.

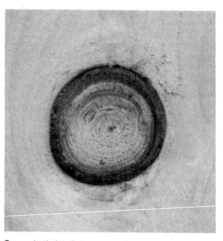
Sound, tight (intergrown), round red knot.

Sound, tight (intergrown), spiked red knot.

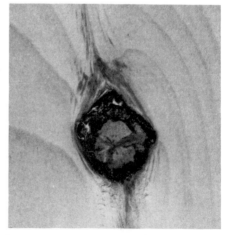
Sound, tight (encased), black knot.

MISSING WOOD: The most common cause is holes, which can be caused by loose knots, worms, bark or pitch pockets.

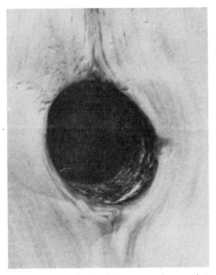

Knot holes: Caused by wood growing around a dead branch.

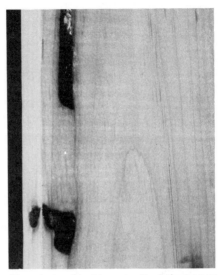

Worm holes: Most insects attack the tree or log just below the bark layer.

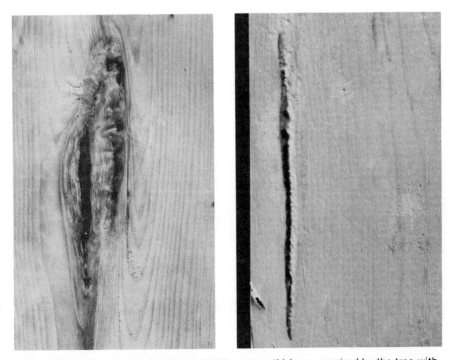

Pockets: Caused by damage to the cambial (tree growth) layer, repaired by the tree with bark or pitch.

Wane: Bark or the lack of wood on the edge of the board—not eased (rounded) at the edges.

CRACKS IN WOOD: Usually caused by dryness.

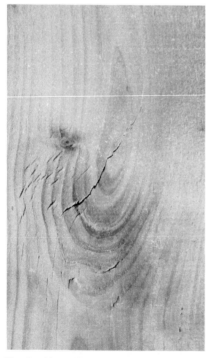

Checks: Normally occur across or through the growth ring.

Shake: Occurs lengthwise between or through the growth ring.

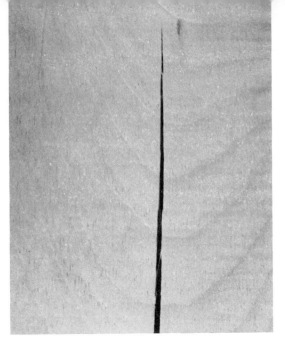

Splits: Occur at the end of a board and go from one side to the other.

DISCOLORATIONS: The marked variation from the natural creamy reddish-brown tints are rated as light, medium, or heavy.

Blue stain: Gray-blue color caused by a fungus which attacks the sapwood before it is fully dry.

Brown stain: Brown color caused by a wood sugar/enzyme reaction to very high temperatures in kiln drying.

Pitch: Honey gold color in streaks and patches; wood may feel slightly sticky.

MANUFACTURING CHARACTERISTICS

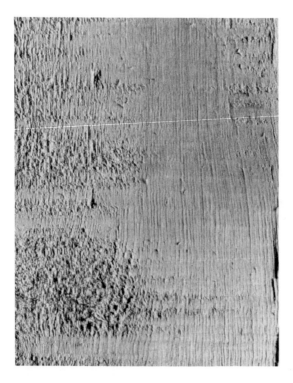

Skip in dressing: Caused by saw deviation; results in a thin rough board in places.

There are positive factors as well, of course. A *bright* board is one with perfect surfaces, unstained and unblemished. A board having a *heart face* is one free of sapwood, pitch pockets, bark streaks, or raised grain.

Before we can go any further with grade classifying, we've got to take a brief refresher course in lumber sizes. Remember, back in Chapter 1, it was noted that *board* has a specific meaning in the lumber vocabulary; a board is any dressed piece less than 2 inches thick and having squared edges. *Dimension lumber* means framing pieces less than 4 inches in the smallest dimension; *timbers* are 5 inches or larger in their smallest dimension. Later, it was mentioned that *finish lumber* is milled boards with exceptionally fine surfaces, and *pattern lumber* has been milled into special shapes. When we come to identifying lumber by grade names, we've got to add a few more.

There is one other manufacturing classification that needs to be noted here, and that's the system used to designate the various ways in which lumber is surface-finished. Actually, these classifications are pretty much academic to anyone except lumber mills, wholesalers, and large-scale builders. Almost every stick of lumber in the average yard or supply store has been surfaced on all four sides. Occasionally, you'll encounter boards that are surfaced only on one side, but this isn't common. At any rate, the classifications are: surfaced four sides (S4S), surfaced two sides (S2S), surfaced one side (S1S), surfaced one edge (S1E), and surfaced two edges (S2E), and there are also combination classifications for edge and side surfacing: S1S1E, S1S2E, S2S1E.

Grades for timbers and dimension lumber begin with *select structural*, the top grade of these two categories of framing lumber. They are followed by the number grades #1, #2, and #3. Small-dimension lumber may also be graded as *construction, standard, utility,* and *stud,* with no numbers being involved.

Boards are either classified as *finish lumber,* the top grade, or as *common,* preceded by a number: #1 common, #2 common, and so on through #5 common. Finish lumber and pattern lumber are called *selects,* and are graded as B and better, C select, and D select. The highest grades quite naturally bring the highest prices in each category.

These grade designations aren't just gobbledygook designed to confuse the average buyer, or to boost the price of the boards he buys. They have a precise meaning, established by grading. The number of natural or manufacturing defects in a piece of lumber of any kind determines its grade, so once you have even a general idea what those grade designations mean, you can decide what grade of lumber you need for anything from a commercial building to a mansion, average house, doghouse, chicken coop, barn, horse corral, or pigsty.

Familiar with grade designations, armed with a knowledge of the difference in appearance and end-use suitability of the different grades of boards, you can order by telephone and be sure that you and the lumberyard salesman on the other end of the line are talking the same language. You'll also be able to look at the lumber that's delivered and know how to sort it out if the pieces of different grades have gotten mixed up when the delivery truck was loaded. You'll know whether the driver unloads your order for B select finish boards with which you're going to panel your living-room wall, or the #3 common that should go to your neighbor down the street who's building himself a chicken coop.

Of course, you'll also have to know how much of what grades to order for whatever job you may have going. Wood undergoes a number of changes in dimensions between the tree and the job, and an inch doesn't always mean an inch in the lumber vocabulary. Let's see how these changes come about, and what they mean to you.

TURNING LOGS INTO BOARDS. When a log makes its first trip through the sawmill, it's reduced to rough-sawn green boards. As a rule of thumb, these measure the *full* number of inches by which they're referred to much later, after seasoning and finishing. A green 1×12-inch board will measure up in those dimensions. The green lumber shrinks during the seasoning process, as its moisture content is reduced to a 19% maximum. In actual practice, air drying will reduce lumber's moisture content to between 12% and 15%, and kiln drying will reduce it to a fairly uniform 15%. This is the minimum amount of moisture that woods will retain in an average climate—if there is such a thing. Or to put it another way, during its life while in use, the moisture content of lumber will vary between 15% and 19%.

Studies made over a long span of years in forest-products laboratories have established 25% as the danger point at which wood decay is most likely to begin. Reducing the moisture content to the 12–15% level is for all practical purposes insurance that during its working life neither air-dried nor kiln-dried wood will under normal circumstances experience an increase in moisture content in excess of 19%. The cellular structure of wood is very much like that of a sponge, which swells a bit when wet, shrinks when dry. This is why doors and windows sometimes prove balky after a long spell of wet weather.

Getting back to our green board, which is now seasoned, and slightly less than 1 inch thick and 12 inches wide, it will measure an average of $25/32$ inch thick and $11^1/_2$ inches wide. If, as is done in some mills handling some types of wood, the board has been dressed while green, these will also be its dimensions after seasoning. Whether finished green or after drying, the board you buy as a 1×12 will measure $3/_4$ inch thick and $11^1/_4$ inches wide. This reduction in size is true of all lumber, with the shrinkage from the *nominal* or *named* dimension being a uniform $1/_2$ inch. In Appendix B you'll find a table that gives these size differences.

All of this, as circumlocutory as it might seem, has a bearing on the way you must figure your lumber order, whether you're buying boards, dimensional lumber, timbers, or finish lumber. Pattern lumber, which includes tongue-and-groove flooring or siding, as well as shiplap and other siding patterns, doesn't follow these rules, remember. We'll look at these types a bit later. What we're talking about now is computing the quantity of boards you'll have to order to cover a given area.

FIGURING QUANTITIES AND PRICES. Let's say you're building a room addition to your house. You'll plan to erect three outside walls, assuming that the present exterior wall will form the fourth wall of the new room. Just as an example, we'll suppose you've decided to build a new room that will be 12×10 feet with an 8-foot ceiling. You arrive at these interior dimensions because they fit neatly into the standard sizes of the 4×8 panelboards you'll use to finish the interior walls.

Figuring the quantity of 2×4 dimensional lumber required is easy. You need one sole plate and two top plates for each wall. The number of studs and rafters will depend on whether you decide to center them 16 or 24 inches apart. You can order these by length, as these are standard stock lengths, and let your dealer figure out how many board feet they contain.

Outside sheathing and siding is another matter. Forget your neat 12×10×8-foot inside dimensions. Your plates and studs will expand the size of your outside walls by approximately $3\frac{1}{2}$ inches at each corner. This means that you'll have to have approximately 14 square feet more sheathing and siding than you have allowed for your inside walls. Add this 14 square feet to the total, subtract the number of square feet that will be left as openings for windows, and perhaps a door as well. This will give you in *square feet* the area your sheathing and siding must cover. Lumber is not sold by lineal feet, though. It's priced by *board feet*, which is a measurement of the *volume* of lumber your purchase represents. Your computations in board feet are done by a standard formula: nominal thickness times nominal width *in inches* times *actual length in feet*. Or:

Nominal thickness × nominal width × actual length (feet) = board feet

Now, convert board feet into square feet—the area to be covered by the sheathing and siding—by using a *conversion factor,* which will be the same for all common-grade boards and dimensional lumber. Here are the factors for boards from 1×4 to 1×12:

4"—1.34   6"—1.20   8"—1.15   10"—1.11   12"—1.10

To use these factors, multiply the area to be covered in *square feet* by the factor given for the width lumber being used. The answer will be in board feet for that width. Then, add 3% to 5% for waste caused by end-cutting which will leave short pieces that are of little or no use. For example, to find the number of board feet of 1×8 common sheathing required to cover an 8×12-foot wall:

$$8 \times 12 = 96 \times 1.15 = 110.4 \times 4\% = 4.416 + 110.4 = 114.816$$

You'd need 115 board feet of the sheathing to cover your wall, from which should be subtracted the number of square feet that window or door openings will take up. In actual practice, you'd subtract these at the beginning, from the square footage.

Certainly, you won't need to do a lot of board-foot computations in figuring your needs for lumber used in craft projects or very small jobs which require only two or three boards. However, knowing how to figure job requirements may save you a substantial sum of money on a large job.

A building axiom is that the wider the nominal width of boards or sidings used in surface coverage, the fewer board feet will be needed. This reflects the character of the board-foot measurement: volume rather than area. So, if you can buy 1×8 boards instead of 1×6 for the same price per thousand board feet, or 1×10 instead of 1×8 for the same price, you'll cover more area for the same cost, and will also have fewer boards to handle, and fewer nails to drive, which will add to the savings. On big jobs, it's certainly to your advantage to figure board-foot coverages and check prices for suitable widths with one or two dealers.

While the conversion factors for common lumber can also be used to convert dimension lumber, they don't apply to pattern lumber such as sidings. Because pattern lumber must be milled to the different shapes in which it is available, extra labor and extra waste is involved in its production. Each of the shapes of pattern lumber—tongue-and-groove, shiplap, bevel siding, bungalow siding, and so on—has its own conversion factor. These shapes are illustrated here, and in Appendix A the conversion factors are given for each one.

SELECTING CONSTRUCTION LUMBER. Now, before moving on to methods of using lumber, let's devote a paragraph or so to the kinds of lumber you'd look for in a major building project, or for a large repair or renovating job.

For foundations, sills, and joists, especially in areas with problems of acid soil, excessive moisture, or termite infestation, your timbers should be of an insect-repellent lumber such as redwood, cypress, or cedar, or of treated lumber such as treated yellow pine.

For studs, rafters, joists, and stringers, use fir, white pine, or yellow pine. If you have long spans, make it yellow pine. If you have extremely

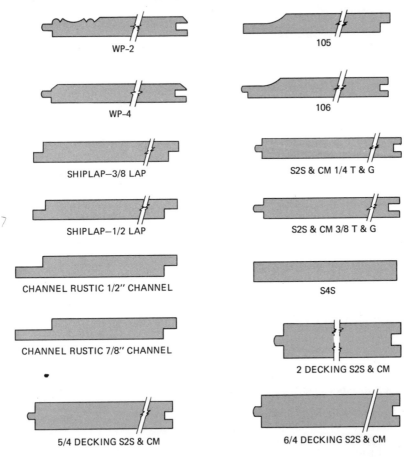

WHITE PINE PATTERN LUMBER

WP-2

105

WP-4

106

SHIPLAP—3/8 LAP

S2S & CM 1/4 T & G

SHIPLAP—1/2 LAP

S2S & CM 3/8 T & G

CHANNEL RUSTIC 1/2" CHANNEL

S4S

CHANNEL RUSTIC 7/8" CHANNEL

2 DECKING S2S & CM

5/4 DECKING S2S & CM

6/4 DECKING S2S & CM

Northeastern Lumber Mfgs. Assn.

These are the shapes of white pine pattern lumber which are generally available. Some are used as exterior siding, some as decking, others as interior paneling or trim.

long spans, give serious consideration to going to one of the hardwoods, such as white oak, or to laminated beams or box beams.

For sheathing and subfloors (except where hardwood floors are to be installed over the subfloor), choose fir, white or yellow pine, or spruce. And of course, for hardwood floors, use white or red oak.

For exterior siding, choose redwood or white pine, with fir or yellow pine as a second choice.

For roofing, choose cedar or redwood shingles or shakes.

One more detail needs to be added at this point. When you buy lumber that isn't going to be used at once, be sure to give it protection

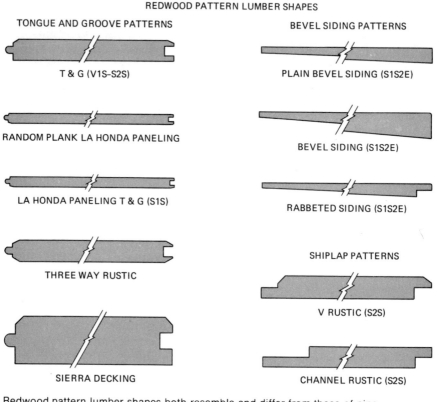

Redwood pattern lumber shapes both resemble and differ from those of pine.

Simpson Timber Co.

while it's awaiting use. Put a couple of timbers on the ground on which to stack the lumber if inside storage isn't available, and cover it with plastic film. Even the best grades of seasoned lumber will absorb moisture from the air in wet weather. It will then swell, and you'll have cracks between boards, loose and creaking subfloors, and even loose nailed joints. Nailing wet lumber is a thankless task, and if you've never tried it, don't.

Protection—preferably an enclosed area—is especially important for moldings and similar trim pieces, as well as for finish flooring and paneling boards. These milled pieces are usually kiln-dried to a moisture content of between 10% and 12% to make them easier to shape in planers and shaping machines. They will pick up atmospheric moisture much faster than will lumber dried to the 15–19% range. During their in-place life, inside a room, the molding and flooring will increase their kiln-dried moisture percentage of 10–12%, and swell. They may increase in size enough to buckle and warp, especially as the nails holding trim are lighter and farther-spaced than those in other types of lumber.

We're getting near the end of our lumber marathon. All that remains is to look at the ways boards are put together to make things: buildings, barns, boxes, benches, bars, baseboards, battens, beehives, belvederes, birdcages, bobsleds, bookcases, braces, and bathtubs, as well as objects that begin with other letters of the alphabet.

However, the foregoing list will provide examples of almost all the ways of joining boards and dimensional lumber together that we need to show. Some of the joints will be useful only in buildings, others will help you put together a bookcase or a box, and the joint that's both the simplest and the most complicated is used in assembling a bathtub. Bathtub? Yes, indeed. That's not just a joke. Wait and see.

NAILS. Nails are the most common fasteners used in joining boards, if you'll forgive such an obvious statement. In building houses, barns, and other structures, nails are indispensable and invaluable. Nevertheless, for small projects, such as cabinets and bookcases and other objects that come under the heading of craftwork, nails are often misused. Many times, they're used alone when they should be used in conjunction with glue, or where a screw would do a better job, or where a brace plus a couple of nails would be more lasting.

At least half the permanence of a nailed joint in lumber lies in choosing the right size and type of nail for the job. Almost 100% of the appearance of an interior in which nails are used depends on the correct choice of nails and their proper use. And close to 100% of the ease of using nails to hold lumber together depends on selecting the right point for the kind of lumber into which the nails are driven.

Nails are sold by weight, and your pound of small nails will have a lot more nails in it than your pound of big ones. The size of nails is expressed in pennies, an inheritance from the British system of selling nails by the hundred several centuries ago. Sixpence bought a hundred nails of one size, tenpence bought a hundred larger nails.

Today, the term is still in use, and when the size of nails is shown in writing, the English abbreviation for penny, the small letter d, is used as a symbol of that size. Thus, an eightpenny nail is written 8d, and so on. From a practical standpoint, the odd-numbered sizes of nails are seldom to be found, except in specialty types, larger than 5d and 7d. Today, nails go by even d designations: 8d, 10d, 12d, and so on up to 16d. At that point, sizes jump in increments of 10; the next size above 16d is 20d, then 30d, and up to 60d.

Any nail smaller than 2d is either a "fine nail" or a brad; any nail larger than 60d is a spike. Brads and fine nails are measured in inches or fractions of an inch rather than by the penny designations. Spikes are measured in inches. The smallest spike is the same length as a 60d nail—6 inches—but the spike is of a heavier gauge wire. Spikes step up 1 inch between sizes, and are designated by their length. The biggest spike is 10 inches long, and the smallest fine nails and brads are $1/2$ inch long. In all nails, brads, and spikes, the diameter or gauge of wire from which they are made increases in ratio to the length.

There are more than two hundred kinds of nails on the market today. Many of them are for special purposes, and we won't be concerned with more than a half dozen of them. In this chapter, the only nails we're interested in are those used in construction and crafts. These include common nails, box nails, finishing and casing nails, roofing nails, and flooring nails. One or two others will be mentioned in passing.

Common nails are the standard by which we'll classify others. Box nails, for instance, have the same shape and point as common nails, but are of smaller diameter in ratio to their length. In nailing thin woods and in nailing close to the ends of boards, box nails are less likely to split the stock than are common nails of equivalent length. (If common nails are used, drill a pilot hole to avoid splitting the boards.)

A nailing rule of thumb when nailing one board to another, face to face, is that the nail used should penetrate the second board within $1/8$ inch of its thickness. This results in a joint that will hold only marginally well, for a 3d nail, for example, isn't very large in diameter, though it is the right length to use when nailing together two $3/4$ inch boards. Use a 3d ring-shank nail, or go to 6d nails cut to a length of $1\,1/4$ inches.

In nailing lumber face to edge, in such applications as installing sheathing or siding, where nails are driven through a board into the edge of a 2×4 stud, a rule of thumb is that the nail should penetrate at least $1\,1/2$ inches into the stud. This calls for a 7d nail. If this size isn't readily available, use an 8d nail, which is only .020 inch greater in diameter. If you're nailing into the edge of $3/4$-inch stock, cut the point of the nail off or use a blunt-pointed hardwood nail to minimize the danger of splitting the edge of the board.

Use standard diamond-point nails on all softwoods except yellow pine and redwood. On yellow pine, use blunt-point nails to minimize splitting. On redwood, you don't have to worry about splitting, but if the nailing is done on exterior work, use a noncorrosive nail—stainless steel, aluminum alloy, or hot-dip galvanized nails—to avoid staining. Moist or wet redwood reacts chemically with exposed ferrous metals, and stains result.

When nailing hardwoods, use a blunt-point nail, or clip the tips off diamond-point nails. On hardwood flooring, use ring-shank or screw-shank flooring nails for best holding power.

For interior nailing on surfaces that will be painted or finished—window sash, door frames, molding, trim—use casing or finishing nails, countersink them, and putty the holes.

FINISHING WOOD SURFACES. Except for redwood, cedar, and cypress, all wood used in exterior applications, where it will be exposed to weather, needs a protective film of paint or varnish to protect it. All three of the woods first named can be stained, painted, or varnished, but they really do just as well if left unfinished. All of them will darken with age, but this doesn't affect their durability.

Today, hardwoods are generally used only in interior applications, and most hardwoods look better when simply varnished. If there is great variation in their tone, it can be adjusted by selective staining before the varnish is applied.

What kind of varnish to use is largely a matter of individual preference. My first choice happens to be tung oil, with urethane a second choice. I prefer to rub the varnish on in two thin coats rather than brushing on one thick coat. The tung oil gives a clear and somewhat less glossy finish than does clear urethane; when using urethane, I like the satin finish. This is, as I said, a matter of individual taste. However, both these finishes reveal the natural grain of the wood in a very attractive manner and are tougher than a hard-rock miner's boots.

Open-pore woods such as oak and walnut need a paste filler rubbed into their grain in order to take varnish smoothly. The filler should be the consistency of a thick batter, and should be rubbed into the wood with a cloth pad in a circular motion. The filler dries quickly, and only a small area at a time can be worked, for excess filler must be wiped off with the grain before it hardens. Use an open-mesh cloth such as cheesecloth in small pads for wiping, and discard each pad when it becomes saturated and stiff.

After the filler, sand with very fine grits, and use a tack cloth—a lintless cloth just dampened with turpentine plus a few drops of clear varnish—to remove dust, grit, and lint. A brush used for varnish should be used only for varnish, and kept scrupulously clean. If the varnish is wiped on, a lintless cloth pad is used. No intermediate or between-coats sanding should be needed, but blobs and bubbles should be smoothed out with a tiny pinch of #6/0 steel wool (some call this grade "steel fur") and the tack cloth used again before a second coat is applied.

Softwoods are usually painted, and as in the case of varnish, the kind of paint used narrows down today to painter's choice. There is very little difference in finish or durability between the alkyd-based and resin-based paints to which today's painter is limited by federal ukase. Neither of these is as versatile or as durable as lead-based paint for outdoor use. Count on getting two to three years of service from any alkyd- or resin-based exterior paint, and don't let the surface go longer than three years unless you live in the mildest kind of climate.

In prime painting—the first coat on raw wood—surface preparation is important. All crevices and cracks should be caulked and painted over.

This nail size chart will help you choose the correct nail for any job. All nails are shown actual size. The lengths of box and finish nails are the same as those of common nails. Finish nails have the same diameters as corresponding sizes of box nails.

Colorado Foundry & Iron Co.

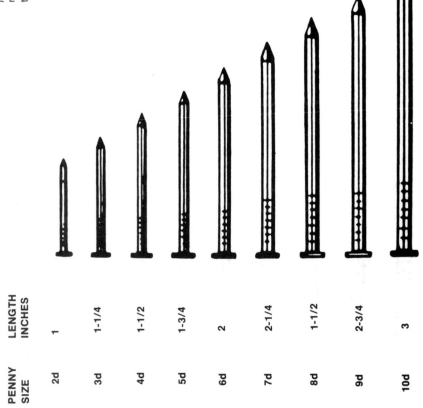

PENNY SIZE	LENGTH INCHES
2d	1
3d	1-1/4
4d	1-1/2
5d	1-3/4
6d	2
7d	2-1/4
8d	1-1/2
9d	2-3/4
10d	3

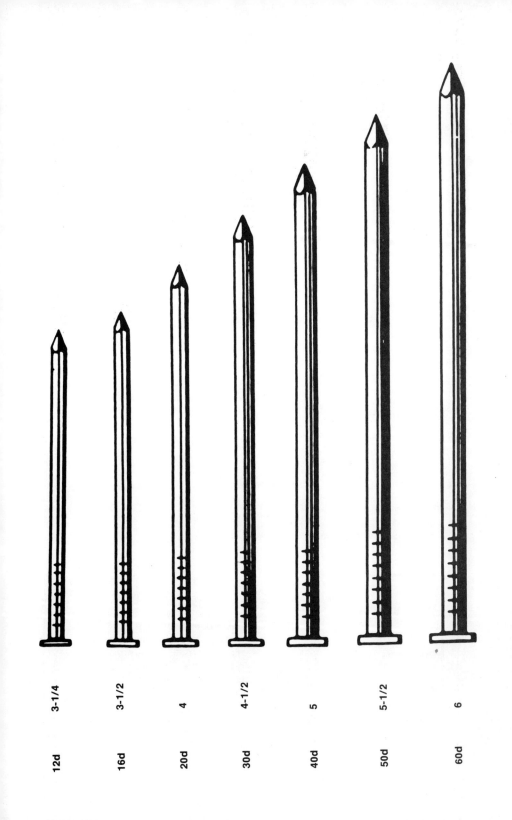

12d   3-1/4

16d   3-1/2

20d   4

30d   4-1/2

40d   5

50d   5-1/2

60d   6

		PENNY SIZE	LENGTH INCHES
LARGE FLAT HEAD			
LARGE FLAT HEAD CHECKERED		2d	1
		3d	$1^1/_4$
FLAT HEAD COUNTERSUNK		4d	$1^1/_2$
		5d	$1^3/_4$
DUPLEX HEAD		6d	2
		7d	$2^1/_4$
OVAL HEAD		8d	$2^1/_2$
		9d	$2^3/_4$
		10d	3
OVAL HEAD COUNTERSUNK		12d	$3^1/_4$
		16d	$3^1/_2$

Nail heads and points vary according to the particular task for which the nail is designed.

	PENNY SIZE	LENGTH INCHES
	20d	4
BLUNT POINT	30d	$4^1/_2$
	40d	5
LONG DIAMOND POINT	50d	$5^1/_2$
	60d	6

NEEDLE POINT

DUCK BILL POINT

CHISEL POINT

These nails are designed to hold better than ordinary smooth-sided nails.

NAILS WITH RINGS   BARBED NAILS   RESIN COATED NAILS

PLASTERBOARD NAILS

ROOFING NAILS

RING SHANK

SCREW SHANK

Plasterboard nails are used in installing gypsum-board interior panels; roofing nails are used in applying vapor barriers such as tar paper or polyethylene sheeting, usually with a "button" cut from cardboard or thin metal to minimize tearing. Screw-shank and ring-shank nails are used when extra holding power is wanted, in flooring installations.

Colorado Foundry & Iron Co.

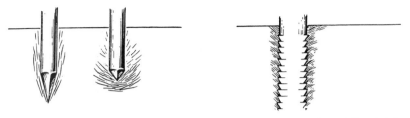

Colorado Foundry & Iron Co.

In this illustration showing the way nails enter wood fibers, the reason why blunt-point nails are less inclined to split a board can be seen. Standard diamond-point nails (left) push the wood fibers aside as they penetrate; blunt-point nails (center) compress the fibers as they go into a board. And the right-hand sketch shows how ring-shank nails lock themselves into the wood fibers and thus hold better.

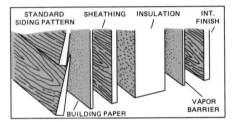

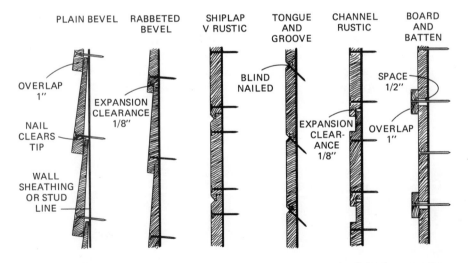

California Redwood Assn.

Here are exterior and interior wall details. At *top* is an exploded view of the layers that make up a standard wall. The *bottom* sketch shows the importance of using the conversion factors when figuring siding needs; the overlaps of all those except the one at right will eat up lumber faster than non-lapped boards.

On old paint being renewed, any loose nails should be pulled and new ones driven to replace them, and all scale and peeling should be removed with steel wool or a fine steel brush and an undercoat applied. For an undercoat on either new or old work, thin the same paint that will be used for the finish coat with 20% of thinners by volume; if a water-based alkyd paint is used, thin with water. Use the paint as it comes from the can for the finish coat. Brush with the grain of the wood in long strokes, working out lap marks to ensure a smooth finish.

Work from the top down to avoid runs that form paint-filled blisters if not picked up. Work an area about 10 feet long from top to bottom, then move on. Most of the lumber now used for exterior siding is very well finished and will take paint readily and smoothly. However, do try to avoid painting on days of high humidity, and also avoid days of blazing heat and bright sunshine, when the paint will dry before you can move on to lap smoothly for the next section. Windy days, if you live in a dry climate, are also bad; the wind will whirl dust and sand onto the wet paint and give it a sandpapery look.

When doing interior painting, you will very probably want to go the undercoater route, especially if your finish coat is to be a semiluster or semigloss or, in busy work areas such as bath, kitchen or utility room, a glossy enamel. The undercoat should be thick—thicker than you think safe. Use a soft brush when applying it to avoid stroke marks. On new work, of course, you will have filled all nail holes with putty and sealed joints with spackle or water putty, and on old work you will have done such scraping and spot sanding as the surface might require.

Remember, whether soft or hard, wood is an organic substance, and all organic matter deteriorates with time. Paint and varnish are your best protection against too-rapid deterioration of a job that's taken you a lot of work to complete.

3 | Choosing and Using Plywood

SINCE THE FIRST successful plywood appeared in 1934, the family has grown enormously. So have its applications. There is now a type of plywood suitable for almost every kind of construction except heavy beams and framing members, and often plywoods can be used to reduce the number of framing members required in a building while still giving the structure the same strength and rigidity that it had with one-fourth to one-third more framing. There is another type of plywood that takes the place of the traditional subfloor and underlayment; tile or roll finish flooring may be laid on it.

Plywood is also available with decorative face designs in hardwood for interior paneling; it can be installed directly to the studs in most cases, saving labor in putting up wall partitions. Traditionally, plywood plays an important part in home craft projects, for it can be joined in large pieces to make deep boxes and cabinets and can be cut with a saber saw into toys that can be used both indoors and outdoors.

Today's plywood mills are a far cry from the early mills, where much of the work was done by hand. Highly specialized machines perform most of the functions of peeling endless sheets of wood from giant logs, rough-cutting them, applying glue, pressing, and trimming. Even the patching of knots in the exterior plies is now done by machines.

UNDERSTANDING PLYWOOD GRADES. Choosing plywood for a specific job is a matter of being able to read the specifications stamped on each sheet. These specifications are uniform, regardless of which of the more than a hundred mills has made it. Once you learn the meanings of the symbols, letters, and numbers of the plywood industry's uniform grade guides, you can choose precisely the kind you need for any project from a building to a delicate jewelry box that will rest on the palm of your hand.

There are thirty-three standard grades of softwood plywood. Thirteen of these are what the industry describes as "engineered" grades, which means that they are designed for structural and construction use in applications where both faces will be covered. The remaining twenty-three grades are described as "appearance" grades, designed to be used in such applications as cabinets, siding, exposed wall paneling, and similar places where one or both sides will be visible. Of the engineered grades, six are for interior use, the others for exterior use. Nine of the appearance grades of plywood are for interior application, the others are for exterior use.

HOW PLYWOOD IS MADE

American Plywood Association.

1. In the first step of plywood manufacture, a knife peels from a rotating log a continuous strip of veneer a little over 8′ wide.

2. Cut into slightly oversized 4×8′ pieces, the veneer is sorted for defects by men standing at a rotating table.

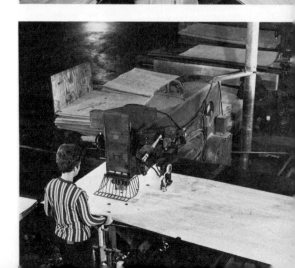

3. Knotholes and other flaws are removed in a punch-press which automatically inserts a patch and glues it in place.

4. Cross-plies precoated with a thermosetting adhesive are put between the long plies.

5. In a laminating press, the plies are bonded by heat and pressure.

6. After cooling, the panels travel through an edger which trims the ends and sides to produce uniform 4×8′ sheets. Other sizes are also made, but the 4×8′ module is the most common.

7. Stacked and banded for shipping, the bundled plywood sheets are moved on pallets to the shipping area.

Most, but not all, grades of plywood are available in oversized sheets up to 12 feet long, though generally these extra-long sheets must be ordered from a distributor or from the mill by local lumber and building-supply retailers. Nor do all retailers stock all thirty-three grades at all times; some of these will have to be ordered if you need them. However, you can almost always substitute with a similar grade that your local dealer carries in stock.

This is also the case with thicknesses. All but a few grades of plywoods are manufactured in four thicknesses, ranging from $^5/_{16}$ to $^3/_4$ inch in the engineered grades and from $^1/_4$ to $^3/_4$ inch in appearance grades. Some of the engineered grades are available in only one or two thicknesses; these are special-application plywoods which would be useless for construction purposes in thin sheets.

In addition to the standard engineered and appearance plywoods there are the specialty varieties intended strictly for decorative and finish use. These plywoods are thinner than the ones intended for use in construction, interior architectural applications, and craft jobs. They are faced with hardwoods and are made for wall paneling or decorative accents or for covering cabinets. They are thicker and somewhat easier to handle than veneers, and are almost always applied with a contact-type adhesive. These plywoods differ from the similar hardboard products in that their facing is of wood rather than being a printed or lithographed design.

Reading plywood grade marks. If building is your business, you will probably carry in your head most of the grade identifications of the various plywoods. If you're a home craftsman or amateur builder, the chances are you'll be familiar with only a few, perhaps none at all. Here, then, are the different grade marks and what they translate into in terms of the characteristics and recommended uses of the plywoods on which they appear.

By way of preliminary, we need first to translate typical markings, such as these:

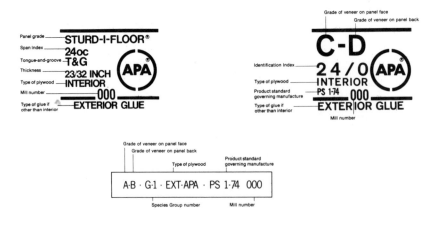

All these marks will have one thing in common. They will carry the letters "APA," which means that the manufacturer has graded the panel in accordance with the industry standards to which members of the American Plywood Association adhere.

Now, the meanings of the letters, figures and words you will see on structural or engineered plywood sheets:

Line 1—Veneer grade or registered trademark or a designation of the plywood's actual use, such as "Underlayment," or letters indicating the veneer grade of the surface or faces.

Line 2—Span index, the maximum recommended spacing for rafters, is indicated by the first figure on the line following the panel grade. If a joist spacing is included, it will be given as a second figure following a slant mark: 24/16; the first figure always indicates rafters, the second always indicates joists. If the edges of a sheet have been specially milled, this will be shown following the span index: "T&G," "Shiplapped," etc.

Line 3—Thickness, on some structural types of sheets.

Line 4—Type, interior or exterior.

Line 5—Product standard. This designation appears *only* on engineered plywoods and refers to the industry standards that govern the sheet's manufacture. This may appear in Arabic numerals and letters or may be given by the word "Group" followed by a Roman numeral.

Line 6—Mill number—the APA number of the mill making the sheet.

Line 7—Type of glue used.

Line 8—National Research Bureau report, if one has been made on this grade of plywood. This will not appear on all types.

On what the trade calls "appearance" plywoods, the markings have many of the same meanings as those on structural grades, but there are enough differences to make it worthwhile to cover the marks line by line:

Line 1—Specific classes or types of wood or the degree of finish of the surfaces or faces of each side of the sheet. The letters used and their meanings are:

N—Smooth-surface natural-finish veneer; select, free of open defects. Allowed are not more than six surface repairs to a 4×8-foot sheet, and these must be inconspicuous, such as oval rather than round plugs, which must match the surface in grain and hue.

A—Smooth, paintable, not more than eighteen repairs.

B—Solid surface, but circular plugs allowed and tight knots up to 1-inch diameter permitted; small splits also permitted.

C—Splits limited to $1/8$-inch width, knotholes or flaws limited to $1/4$ inch square; wood-paste repairs permitted; minor broken grain permitted.

C—Tight knots to 1½ inches permitted, also knotholes 1 inch across the grain. Paste and plugs permitted; also off-color areas or sanding defects that do not impair strength.

D—Knotholes and knots to 1¼-inches width across grain permitted; limited splits allowed.

Line 2—Standards group under which sheet was manufactured. (If a span index is required—and not all plywood of this type has this—it will appear as line 3.)

Line 4—Product standard number under which sheet was made.

Line 5—APA number of the manufacturing mill.

Not all plywood sheets carry edge-grade marks. When used, the edge marks simply condense the basic information required.

Now, here's the entire range of plywood grades by grade mark, recommended use, outer-veneer types, and thicknesses available.

 Use as wall and roof sheathing, subflooring. Thicknesses: ⁵⁄₁₆, ³⁄₈, ½, ⁵⁄₈, ³⁄₄.

 Same uses as (1), but this grade made with exterior glue. Same thicknesses as (1).

 Unsanded, for structural use: box beams, gusset plates, stressed skin panels, containers. Thicknesses: ⁵⁄₁₆, ³⁄₈, ½, ⁵⁄₈, ³⁄₄.

 Combination subfloor-underlayment. Surfaced to resist impacts, and to accept installation of resilient finish materials: tile, roll vinyl, etc. Available square-edge or T & G. Thicknesses: ¹⁹⁄₃₂, ⁵⁄₈, ²³⁄₃₂, ³⁄₄.

 Combination subfloor-underlayment for spans 32 to 48 inches. Impact-resistant surface accepts resilient floor finish materials. Available square-edge or T&G. Thickness: 1⅛ only.

 For use over structural subfloor, for application of resilient finish materials. Also available with exterior glue. Thicknesses: ¼, ³⁄₈, ½, ¹⁹⁄₃₂, ⁵⁄₈, ²³⁄₃₂, ³⁄₄.

For built-ins, wall-tile and ceiling-tile backing, walkways, separator boards. Not a substitute for (6); surface not impact-resistant. Thicknesses: $^1/_4$, $^3/_8$, $^1/_2$, $^{19}/_{32}$, $^5/_8$, $^{25}/_{32}$, $^3/_4$.

 For subflooring and roof decking, siding on farm buildings, treated-wood foundations. Unsanded with waterproof bond. Thicknesses: $^5/_{16}$, $^3/_8$, $^1/_2$, $^5/_8$, $^3/_4$.

 For engineered construction and industrial applications requiring full exterior-type panels. Unsanded. Thicknesses: $^5/_{16}$, $^3/_8$, $^1/_2$, $^5/_8$, $^3/_4$.

 Combination subfloor-underlayment. Same as (4) and (5) except made with exterior-grade glue for high moisture resistance. Thicknesses: $^{19}/_{32}$, $^5/_8$, $^{23}/_{32}$, $^3/_4$.

 For use over structural subfloor. Same as (6), except made with waterproof glue. Thicknesses: $^1/_4$, $^3/_8$, $^1/_2$, $^{19}/_{32}$, $^5/_8$, $^{23}/_{32}$, $^3/_4$.

 For use under severe moisture conditions: tile backing in refrigerated storage areas, controlled-atmosphere rooms, tanks, boxcar and truck floors, open soffits. Thicknesses: $^1/_4$, $^3/_8$, $^1/_2$, $^{19}/_{32}$, $^5/_8$, $^{23}/_{32}$, $^3/_4$.

 Reusable concrete-form plywood. Mill-oiled. Thicknesses: $^5/_8$, $^3/_4$.

NN G1 INT APA PS174 000
NA G2 INT APA PS174 000
Cabinet-quality with natural finish both or one side. Special-order item. Thickness: $^3/_4$ only.

ND G2 INT APA PS174 000
Cabinet-quality, natural finish one side only. Special-order item. Thickness: $^1/_4$ only.

AA G1 INT APA PS174 000
Cabinet-quality with both faces smooth, for built-ins, cabinets, furniture, partitions. Thicknesses: $^1/_4$, $^3/_8$, $^1/_2$, $^5/_8$, $^3/_4$.

AD G1 INT APA PS174 000 — Same as (16), but with only one smoothed face for painting or varnishing. Thicknesses: $^1/_4$, $^3/_8$, $^1/_2$, $^5/_8$, $^3/_4$.

A-D GROUP 1 INTERIOR APA 000 — For use where appearance of only one side is important. For paneling, built-ins, shelving. Thicknesses: $^1/_4$, $^3/_8$, $^1/_2$, $^5/_8$, $^3/_4$.

BD G2 INT APA PS174 000 — Utility paneling with two solid sides, but grade permits circular plugs in face veneer. Thicknesses: $^1/_4$, $^3/_8$, $^1/_2$, $^5/_8$, $^3/_4$.

B-D GROUP 2 INTERIOR APA 000 — Utility panelling, same as (19) but with only one solid side. Thicknesses: $^1/_4$, $^3/_8$, $^1/_2$, $^5/_8$, $^3/_4$.

DECORATIVE BD G1 INT APA PS174 — For interior wall finish; available with face rough-sawn, brushed, grooved, or striated. For use as accent wall, facing for built-ins, displays, exhibits. Thicknesses: $^5/_{16}$, $^3/_8$, $^1/_2$, $^5/_8$.

PLYRON INT APA 000 — Faced on both sides with hardboard instead of wood veneer. For countertops, shelving, cabinet doors, flooring. Faces available: tempered, untempered, smooth, or screened. Thicknesses: $^1/_2$, $^5/_8$, $^3/_4$.

AA G1 EXT APA PS174 000 — Finished both sides for painting or varnishing. For cabinets, boats, built-ins, signs, ducts, tanks, fences. Thicknesses: $^1/_4$, $^3/_8$, $^1/_2$, $^5/_8$, $^3/_4$.

AB G1 EXT APA PS174 000 — Same as (23), but finished one side. Thicknesses: $^1/_4$, $^3/_8$, $^1/_2$, $^5/_8$, $^3/_4$.

A-C GROUP 1 EXTERIOR APA 000 — For use where appearance of only one side is important: fences, soffits, farm buildings, tanks, trays, structural uses. Thicknesses: $^1/_4$, $^3/_8$, $^1/_2$, $^5/_8$, $^3/_4$.

BB G2 EXT APA PS174 000 — Utility panel with solid faces. Thicknesses: $^1/_4$, $^3/_8$, $^1/_2$, $^5/_8$, $^3/_4$.

B-C
GROUP 2 (APA)
EXTERIOR
PS 1 000

Utility panel; for use as base for exterior coatings for walls, roofs, farm service, work buildings, storage sheds, truck lining, tnaks. Thicknesses: $^1/_4$, $^3/_8$, $^1/_2$, $^5/_8$, $^3/_4$.

HDO 60 60 BB PLYFORM EXT APA PS174

High-density overlay paneling. Has hard, semiopaque resin-fiber coating on both faces. High resistance to abrasion. For countertops, work surfaces, signs, tanks, cabinets, concrete forms. Thicknesses: $^3/_8$, $^1/_2$, $^5/_8$, $^3/_4$.

MDO BB G2 EXT APA PS174 000

Medium-density overlay paneling. Has smooth, opaque resin-fiber coating on one or both faces. For siding and similar outdoor applications, built-ins, signs, displays. Excellent paint base on coated side. Thicknesses: $^3/_8$, $^1/_2$, $^5/_8$, $^3/_4$.

303 SIDING   6 S
GROUP 1
24 oc SPAN (APA)
EXTERIOR
PS 1 N 000

Proprietary (Simpson Timber Co.) paneling for exterior applications such as siding, fencing, etc. Surfaces include V-groove, channel-groove, striated, brushed, rough-sawn, texture-embossed with medium-density overlay. Thicknesses: $^3/_8$, $^1/_2$, $^5/_8$.

303 SIDING   6 S W
T 1 11
GROUP 1
16 oc SPAN (APA)
EXTERIOR
PS 1 000

Proprietary (Simpson Timber Co.) paneling for exterior siding. Edges shiplapped. Grooved, $^1/_4$ inch deep, $^3/_8$ inch wide, on either 4-inch or 8-inch centers. Available unsanded, textured, and with medium-density overlay. Thicknesses: $^{19}/_{32}$, $^5/_8$.

PLYRON EXT APA 000

Hardboard-faced on both sides; available with tempered hardboard, smooth or screened. For countertops, shelving, cabinets, doors, flooring, in moisture-prone areas. Thicknesses: $^1/_2$, $^5/_8$, $^3/_4$.

MARINE AA EXT APA PS174 000

Marine plywood, for boat hulls and boat fittings. Made only from Douglas fir or western larch. Special jointed-core construction. Available with high- or medium-density-overlay faces; see (28) and (29). Thicknesses: $^1/_4$, $^3/_8$, $^1/_2$, $^5/_8$, $^3/_4$.

By referring to the descriptions you can choose precisely the grade of plywood you require for any job where you plan to use it, and know just what to order from your retail dealer. You'll see that in some cases you can save a lot of work and get a better job done faster by selecting a plywood such as the medium-density-overlay panels, which give you a good painting base for exterior jobs, and save you both the cost and time of a prime coat. This is also true of several of the interior grades.

USING ENGINEERED PLYWOOD. Even when you're putting up a building, such as a garage, or building an addition to a house, there will be times when plywood, used in company with lumber, will save you a great deal of time. A typical example would be in fabricating a load-bearing beam to give you the long open span required to accommodate a garage door, or a glass window wall.

Your first step is to make a box the size of the beam. In the sketch, a 20-foot beam is shown as an example of this kind of construction. Two 20-foot lengths of 2×4 lumber form the top and bottom, two lengths of 2×4s make the end pieces, and single 2×4 stiffeners are nailed at 4-foot intervals along the interior. The box is then covered end to end and edge to edge with $^3/_4$-inch structural plywood, grade $^{32}/_{16}$, nailed to each of the top and bottom 2×4s with 8d nails spaced 3 to $3^1/_2$ inches apart. Similar double rows of nails are driven into each of the two 2×4s of frame ends, and a single row of nails is driven into each stringer. Finally, 8-foot splices made of $^1/_2$-inch plywood are centered on the beam on both sides and nailed. Box beams of 12, 14, 16, 18, or 20 feet—even longer, if

Box beams save weight and cost when used across long spans, such as those required for double or triple garage doors.

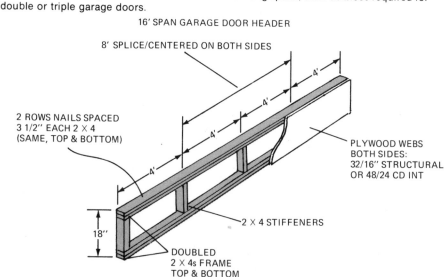

16′ SPAN GARAGE DOOR HEADER

8′ SPLICE/CENTERED ON BOTH SIDES

2 ROWS NAILS SPACED 3 1/2″ EACH 2 X 4 (SAME, TOP & BOTTOM)

PLYWOOD WEBS BOTH SIDES: 32/16″ STRUCTURAL OR 48/24 CD INT

2 X 4 STIFFENERS

18″

DOUBLED 2 X 4s FRAME TOP & BOTTOM

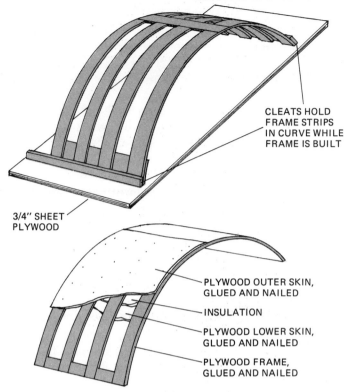

CLEATS HOLD
FRAME STRIPS
IN CURVE WHILE
FRAME IS BUILT

3/4" SHEET
PLYWOOD

PLYWOOD OUTER SKIN,
GLUED AND NAILED

INSULATION

PLYWOOD LOWER SKIN,
GLUED AND NAILED

PLYWOOD FRAME,
GLUED AND NAILED

Plywood enables you to use curved structural accents such as an arched entryway. Use exterior-grade plywood and waterproof glue—Cascamite or equivalent—to build up the frame between cleats nailed to a sheet of plywood. Apply the skin with glue and nails, using clamps to hold skin and frame together until the glue sets up.

Gussets of structural plywood give added strength to roof trusses, especially those that must be cantilevered to provide the overhang for an entryway or at the eaves.

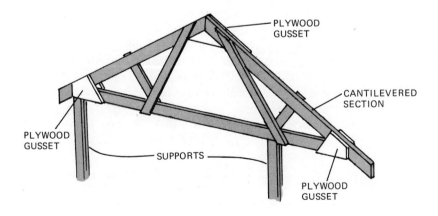

PLYWOOD
GUSSET

CANTILEVERED
SECTION

PLYWOOD
GUSSET

SUPPORTS

PLYWOOD
GUSSET

required—are stronger and more rigid per pound of weight than solid-timber beams, and much less costly.

Plywood gussets can be used to brace trussed roof joints, as shown in the next sketch, and will materially increase their strength over gussets made from boards. If you want an arched entryway or a loggia, you can make curved panels from plywood as shown in the third sketch. These are only a few of the ways in which plywood and lumber can be combined to give you a structure that has superior internal construction features.

For exterior finishing or remodeling, plywood panels can also be very helpful. Panels made for use as exterior siding enable you to remodel an older house to present almost any appearance you desire: sleek modern with painted medium-density overlay; rustic with rough-sawn stained panels; contemporary with striated painted panels; woodsy with stained brushed panels; traditional with V-groove or channel groove either stained or painted. The panels can be applied to new or old sheathing. They can be used with insulation batts that incorporate a vapor barrier, over old siding in a remodeling job, by using furring strips.

One of the benefits that plywood siding offers the do-it-yourself remodeler is the speed with which a wall can be covered, even by a lone worker. There are also many options in fitting the panels. They can be applied horizontally or vertically, and no special tools are required, nor any special skill beyond the ability to swing a hammer and guide a saw in straight cuts. The vertical and horizontal joints shown in the sketches will give you an idea of what to expect if you undertake a job of either building from scratch or remodeling by covering existing siding.

A problem that many nonprofessional remodelers have in putting up board siding is making weathertight corner joints. Plywood siding solves corner problems in ways that even an amateur carpenter with two left hands, each having five thumbs, can solve without sweat or tears. The sketches give details. Window and door openings are worked out with equally simple solutions.

There is really only one precaution that must be taken when installing exterior plywood siding. Each edge of each panel must be protected. This is the area of vulnerability to water damage on plywood siding, even that graded for exterior use.

While the plywood makers don't recommend any specific protection other than the simple "caulking," here's my own based-on-experience suggestion. Use the best grade of latex caulking compound. Apply a $1/2$-inch line to the studding or furring strip, then use a *wet* putty knife to spread the caulking compound into a thin, even coat.

Do this immediately before installing the panel, and remember that your putty knife must be kept wet at all times. Have a small can of water within easy reach and dip the putty knife into it often. Then, apply a $1/8$-inch line of caulk around the perimeter of the panel and spread it the

Typical textures available in plywood decorative panels for exterior or inside use. Different manufacturers offer different versions of textured panels, as well as plain woodgrains.

Roughsawn,

Plain V-groove,

Channel Groove,

Brushed Board
and Batten,

Weathered
Board and Batten.

Striated.

same way. Place the panel and nail it at once. Repeat the procedure as you install each panel.

By using this technique, you eliminate the danger of voids in the caulking and will have a completely weathertight seal. If you have a small oozing of caulk when you butt a panel to one already in place, be sure to wipe the caulk away at once with a wet cloth. If you fail to do this, the caulk will prevent either stain or paint from adhering.

USING INTERIOR PLYWOOD. When we move indoors, applications of plywood to new construction or remodeling multiply rapidly. They are also very numerous in projects usually called home improvements or modernizing. For all these jobs, you can consider using plywood alone, or in conjunction with other materials, primarily surface finishes such as wood veneers and laminates. Since these will be discussed in a later chapter, we will skip them for now.

Tools and working methods. Virtually every interior project you take on will involve finishing plywood in one manner or another. The way to make the final steps easier and simpler is to bear in mind from the start of your project that there's finishing work ahead of you. This means handling the plywood sheets carefully during the steps required to reduce them to the pieces that will be assembled into your project as the job progresses.

Dents, mars and gouges, tool marks, and the results of miscalculated cuts are easier to prevent than to cure. The face veneers of all plywoods except the high-density-overlay (HDO) and medium-density-overlay (MDO) grades are, in comparison to solid-wood surfaces, relatively delicate. Deep scratches, dents, and torn wood fibers, can be avoided by proper storage and careful handling. Avoid bouncing sheets on their corners and edges. Be sure they are stored in a way that will give maximum protection to the sides that will be exposed in the finished job. Stacking plywood with the finish face down on a gravel driveway is an extreme example of mishandling, but it's one I've encountered.

Plywood lends itself to working with the simplest hand tools as well as power tools. A crosscut saw is better than a ripsaw; a fine-cut power-saw blade is better than a combination blade. For very fine cuts with saber saws, a metal-cutting blade can be a great splinter-saver. You'll have to advance the saw a bit slower than you would a coarse blade, but the time lost is small when compared to the time you'd spend smoothing a cut made with a blade coarse enough to splinter the edges.

Other hand tools that will be helpful are a keyhole saw for radius cuts, a block plane for smoothing edges, and a half-round mill bastard file for smoothing edges after planing. Sandpaper used should be of the open-face type, as plywood glues will clog closed-faced papers.

Your three basic power tools are a circular saw for long, straight cuts, a saber saw for radius cuts, and a power sander; this will replace both plane and file in finishing edges. If you're going to make joints, a router is handy, but you can do most cabinet joinery with rabbet end cuts for joining corners and dado cuts for assembling shelves or the equivalent of shelves. Both these can be made with a circular saw by raising its blade

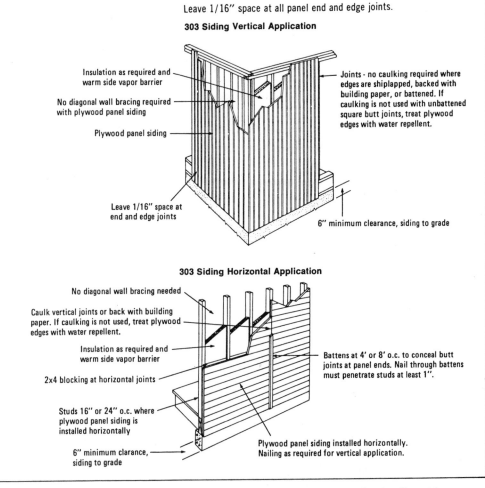

Leave 1/16" space at all panel end and edge joints.

303 Siding Vertical Application

Insulation as required and warm side vapor barrier

No diagonal wall bracing required with plywood panel siding

Plywood panel siding

Leave 1/16" space at end and edge joints

Joints - no caulking required where edges are shiplapped, backed with building paper, or battened. If caulking is not used with unbattened square butt joints, treat plywood edges with water repellent.

6" minimum clearance, siding to grade

303 Siding Horizontal Application

No diagonal wall bracing needed

Caulk vertical joints or back with building paper. If caulking is not used, treat plywood edges with water repellent.

Insulation as required and warm side vapor barrier

2x4 blocking at horizontal joints

Studs 16" or 24" o.c. where plywood panel siding is installed horizontally

6" minimum clarance, siding to grade

Battens at 4' or 8' o.c. to conceal butt joints at panel ends. Nail through battens must penetrate studs at least 1".

Plywood panel siding installed horizontally. Nailing as required for vertical application.

American Plywood Assn.

Installation details above show you how to apply both vertical and horizontal exterior plywood siding panels.

to give you a cut of the desired depth and clearing excess wood with a chisel. See the sketches for details.

In sawing with a handsaw, the face or finish surface of the plywood sheet is always up. In sawing with a power saw, the face is down. If a coarse blade is used in the saw, apply strips of masking tape along the cutlines on both sides to minimize splintering.

When filing or planing, work from the finish side of the sheet when truing or smoothing edges. If any splintering takes place, it will be on the edge that will be hidden. Use long, smooth forward strokes with both

Plywood Siding Joint Details

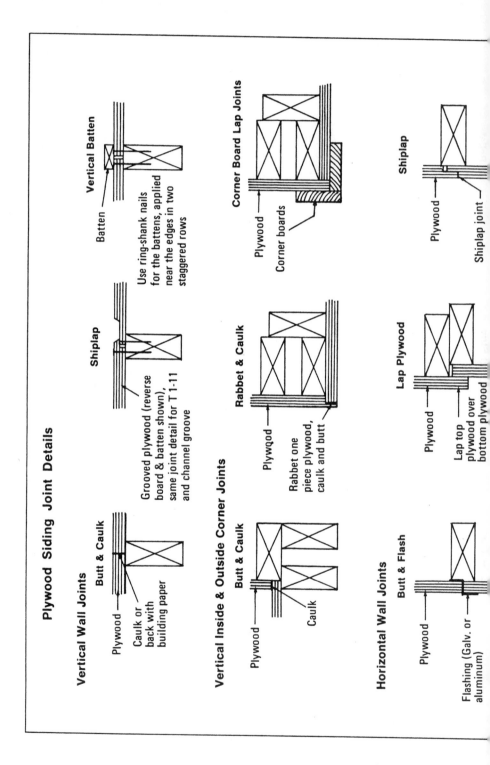

Vertical Wall Joints

Butt & Caulk

- Plywood
- Caulk or back with building paper

Shiplap

Grooved plywood (reverse board & batten shown), same joint detail for T 1-11 and channel groove

Vertical Batten

- Batten

Use ring-shank nails for the battens, applied near the edges in two staggered rows

Vertical Inside & Outside Corner Joints

Butt & Caulk

- Plywood
- Caulk

Rabbet & Caulk

- Plywqod

Rabbet one piece plywood, caulk and butt

Corner Board Lap Joints

- Plywood
- Corner boards

Horizontal Wall Joints

Butt & Flash

- Plywood
- Flashing (Galv. or aluminum)

Lap Plywood

- Plywood
- Lap top plywood over bottom plywood

Shiplap

- Plywood
- Shiplap joint

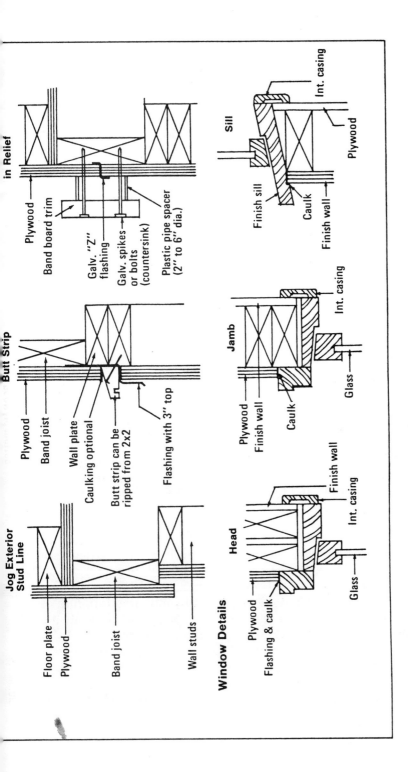

in Relief

Plywood
Band board trim

Galv. "Z" flashing
Galv. spikes or bolts (countersink)
Plastic pipe spacer (2" to 6" dia.)

Sill

Int. casing
Plywood
Finish sill
Caulk
Finish wall

Butt Strip

Plywood
Band joist
Wall plate
Caulking optional
Butt strip can be ripped from 2x2
Flashing with 3" top

Jamb

Int. casing
Plywood
Finish wall
Caulk
Glass

Jog Exterior Stud Line

Floor plate
Plywood
Band joist
Wall studs

Window Details

Head

Plywood
Flashing & caulk
Finish wall
Int. casing
Glass

American Plywood Assn.

All the joint details needed to guide you in proper, weather proof application of exterior plywood siding panels are shown above.

plane and file, never a back-and-forth stroke. The latter is apt to cause splintered edges on the finish side of the sheet.

In power sanding of surfaces to be finished, the coarsest abrasive grades should be avoided. A #120 or even #150 open-faced garnet production paper is the one I've found best to use in a power sander, with a #220 grade of the same paper for a fine finish.

Always use glue in addition to nails or screws to hold 90° joints. Look on the nails or screws as merely devices to hold the joint tight while the glue sets up. Be sure to wipe up glue spills or squeezed-out threads quickly. If the glue is allowed to penetrate a surface that is to be stained or varnished, it will seal the surface against the final finish.

Uses of interior-grade plywood. For openers, let's consider the purely decorative aspects of interior plywood. The first and most obvious is creating an accent wall of natural wood with hardwood-veneered plywoods. You can apply the veneer yourself, if you wish. Using the new ultrathin veneer available in rolls, this job is no more difficult than putting on wallpaper or contact paper. The working method is the same. These veneers are backed with paper and are easy to apply, using the newest contact cements, which allow you to shift the veneer for aligning before rolling it into contact and final adhesion.

There are thin hardwood-veneered plywood panels designed for application to existing walls. These aren't covered in our grade guide, but most retail lumber and building supply outlets carry them in stock. These decorative panels come in standard 4×8 sheets, are $3/_{32}$ or $1/_8$ inch thick, and are applied with a mastic. Brads or fine nails with heads that match the wood are available for securing edges and placing moldings. The only tools required to install these sheets are a hammer or mallet and a padded wood block. A saw will be needed as well to cut the sheets to size as required and to make cutouts around door and window openings.

Another way to create an accent wall is to apply one of the textured plywoods, either in half-sheets as wainscoting or in full sheets as individual panels, and either covering the wall or alternating with smooth-wall areas. The textured plywoods—striated, brushed, or patterned—can be stained, painted, or varnished. Or they can be "antiqued," using the same technique employed in antiquing furniture. A saw, hammer, and nail set are the only tools required to install the plywood and the finish moldings. Because the lower areas of walls in busy rooms tend to get a bit scarred and battered, and because plywood provides a much more durable surface than either plaster or paper, wainscoting is an exceedingly popular renovating job.

Soffit or shielded lighting, concealed behind a shallow rim of plywood circling the room, is another easily achieved decorative effect. Cut a narrow strip of $^3/_4$-inch plywood to serve as a base for the light-shielding strip and as a means of mounting sockets for the small, low-wattage bulbs that are used. The base can be installed by screwing it to the ceiling joists; it will be hidden, as will the lights, by the shield. The base can be parallel to the perimeter walls of the room, or can be curved in at the corners. Attach the shield of $^1/_4$-inch plywood, cut in strips, to the slanted edges of the interior perimeter of the base; the sketch shows working details. A dimmer switch to control the intensity of the lighting is a nice added touch.

If you checked the listing of plywood grades a few pages back, you'll have noticed that there are engineered grades of plywood designed for floor underlayment. These are made with a surface which is impact-resistant and load-resistant, and which is also smooth enough to allow the installation of such popular resilient floor coverings as thin vinyl tiles and roll vinyls directly over the plywood. These underlayment sheets can cure many a problem floor, but if the problem happens to be one of a badly buckled subfloor, it might be the course of wisdom to strip the room down to the floor joists and install one of the grades of plywood which serve as both subfloor and underlay.

When making this kind of installation, in either new construction or remodeling, be sure to follow the manufacturer's recommendations and use the proper size of ring-shank nails to ensure that the floor will stay tight and squeak-free. You might even look at the prospects of using the glue method of handling these combination subfloor-underlayment jobs. Gluing makes a single unit of the joists and subflooring, and ends both creaking and sagging. Both the structure of the house and the floor are stronger.

Interior cabinets. Cabinetwork of all kinds is easy to carry out with plywood. Use A-A INT or A-B INT grade depending on whether one or both faces of the plywood will be exposed. Or use N-N grade, if you plan to varnish the exposed surfaces and let the natural woodgrain add a special decorative touch. The A-grade surfaces also accept varnish well, and both N and A grades are designed to make painting easy and to avoid extensive surface preparation such as sanding.

Music walls, which provide planned places for TV, stereo, tapes and casettes, records, TV game devices, and movie or slide projectors and screens, come to mind at once when interior cabinets are up for discussion. So do decorative cabinets to be used for additional storage or for the display of collections or hobby items. Bookcases are another type of installation easy to construct of warp-free plywood.

In the kitchen, a built-in dinette with foldaway table and bench seating can make the best use of limited space and make the kitchen roomier for meal preparation. Often, kitchen efficiency can be improved and labor saved by building a work island in the room's center. Extra drawers and cabinets can provide more storage facilities and end clutter. For that matter, new cabinets of plywood—in the kitchen or any other room—can be installed to ease storage problems. If floor space is scarce, the cabinets might be wall-mounted. Attach them to studs with screws going through the cabinet's back, or if the wall is of concrete block or masonry, put them up with toggle bolts.

It isn't the purpose of this book to provide detailed working plans for your projects, but rather to acquaint you with the basic methods of using the various materials. These basics include the tools needed, and how to make the key joints that have proved easiest to work with and most satisfactory for each of the materials discussed. In Appendix D you'll find a list of trade associations and manufacturers, all of whom have free literature that contains how-to-do-it plans for specific projects. Write them and ask for plans, if you don't feel that you can go ahead without detailed drawings. However, using the data given in the preceding and following pages, you should be able to make your own drawings for your own special needs.

Finishing plywood surfaces. When we say "finishing" we automatically think first of paint, stain, and varnish. That's not where we're going to start this portion, though, because a large number of home craftsmen seem to have greater problems in finishing off plywood edges than in finishing surfaces. Admittedly, the cross-grained nature of these edges do present a challenge. You can't just ignore them and hope they'll go away.

There are half a dozen ways to finish off a plywood cabinet project. In work areas such as kitchen, bathroom, and laundry room, where a majority of work surfaces are covered with vinyl or laminate, the most generally accepted solution is to use a metal crown strip. These are trim strips shaped like an inverted capital L, with the short side finished to pull down on the surface and the long side drilled and countersunk to accept oval-head or round-head screws. The strips can be bent into a relatively sharp curve, and corners can be mitered very easily with the corner of a straight file or a triangular file.

Metal doesn't look too good in other rooms, though. Here wooden moldings should be used. There are five stock moldings which virtually every lumberyard and building-supply house has on hand in assorted dimensions which will finish off edges of shelves and cabinets or other built-ins. These are beveled strip, half-round, quarter-round, base-shoe, and cove. They are reproduced in profile in the sketch, and the way to miter them at corners is shown in the following sketches. These wooden moldings can be painted, stained, or varnished to match whichever finish you select.

Veneer strips, most of them self-adhering, are also available in both hardwood and softwood varieties to match any fine-wood finish. These

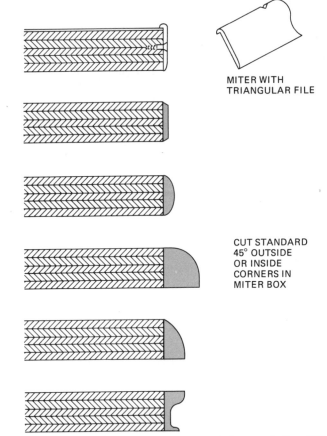

MITER WITH
TRIANGULAR FILE

CUT STANDARD
45° OUTSIDE
OR INSIDE
CORNERS IN
MITER BOX

Methods of finishing the edges of plywood countertops, shelves, and other interior jobs. All except the core can be curved around a moderate arc.

thin strips are generally $3/4$ inch wide, but they can be trimmed after installation with a craft knife or razor saw, using the bottom edge of the plywood as a cutting guide.

Finally, you can sand the edges of plywood sheets, apply a thin coat of filler to them, give the dried filler a quick rubdown with #400 wet-or-dry silicone-carbide auto-body paper used wet, and have as nice a surface as you could ever want to paint on.

As for general surface finishing, if you've done your advance studying of plywood grades with enough attention to detail, you will have eased the job of giving your plywood jobs a suitable finish by at least 50% to 75%. My own rule of thumb when working with plywood is to choose the least expensive grade practical in the application for which it will be used. In practice, this means selecting an N grade for any surface that is to be stained or varnished, or varnished after staining, or waxed. An A or B grade is chosen for exposed surfaces which will be painted, and a C or

D grade for interior, hidden surfaces. This reduces very substantially the amount of time which must be spent in such pre-painting preparations as sanding and filling.

However, if you're building a cabinet or other semi-furniture piece from a plywood faced with an open-grain hardwood such as oak or walnut, you must be prepared to follow the same steps that are used in finishing fine furniture. This begins with sanding, goes on to rubbing in a paste filler of the proper shade to match the wood, and resanding, then varnishing with a good rubdown, using felt pads and a tack cloth (see Chapter 2) between varnish coats.

On any plywood surface that is to be painted, a decent minimum of sanding will be needed. This can be done by hand, using a sanding block, or with a power sander. If the surface is to be varnished, either with or without having been stained, or if a high-gloss lacquer or enamel is going to be applied rather than a gloss or semigloss paint, a sealer is very important. On painted surfaces, using the filler eliminates the need for a filler-type undercoat, which saves both paint and labor.

A fairly thin undercoat on a sanded and sealed surface of virtually all A-grade and most B-grade plywoods should give you a finish equivalent to that which you'd get with a single coat on a factory-prepared surface such as N or A.

To repeat an analogy which I've used before in a different context, a fine wood surface resembles a pretty woman's face in one respect. Left unadorned and unprotected, it's attractive, even beautiful, but it needs a bit of help from makeup to bring out its best features and to glow with loveliness. Paint and varnish are the only cosmetics in wood's makeup kit. Don't fail to use them judiciously to help wooden surfaces look their best.

4 | Choosing and Using Composition Boards

FOR THE SAKE of brevity and clarity, the term "composition boards" is used here to describe four different products. In the terms of the industries which produce them, these are called *particleboard, medium-density fiberboard, hardboard,* and *softboard* or *pulpboard.* We'll deal with the first two as though they were one, because their characteristics are quite similar and the methods of making them virtually the same.

To distinguish between these products at the very start:

Particleboard and *medium-density fiberboard (MDF)* are hard-surfaced panels made from flakes of wood, splinters, wood chips, and selected screened mill wastes or by-products such as sawdust. The solids are mixed together in a liquid bonding agent and pressed into panels.

Hardboard is made from wood pulp processed to reduce it to very fine filaments and bonded with a liquid under heat and pressure into thin, lightweight panels which are flexible rather than rigid, as are the particleboards.

Softboard is made from highly processed wood pulp that has been reduced with chemicals and heat into small grains. The grains are pressure-shaped with a bonding agent into panels or tiles, prefinished on one side. These are also called *pulpboards.*

Within these broad general categories there are some variations, which will be covered in detail as each group is discussed. Each of these composition boards is fabricated into general-utility products designed to give the maximum service to the average user, and each is also made in special custom forms to suit the specific needs of large-scale users, such as furniture manufacturers, electronics manufacturers, and so on. We won't be going into these custom applications here, but will occasionally touch on them in passing.

PARTICLEBOARD AND MDF. Particleboard in its modern form is a relatively new product, and its full range of uses is still being explored. Being less expensive than either hardboard or plywood, particleboard has come to be very widely used as a floor underlayment, as subflooring, and as cores for built-ins and free-standing furniture pieces which are covered by laminates or veneers.

Generally, particleboard will cost from one-fourth to one-third less than plywood, and though its weight generally precludes it being used for doors, it is widely used in place of plywood as a substrate or foundation material for vinyl or laminate-covered countertops in kitchens and bathrooms. It is also produced in special forms for such applications as stair treads, where it is used in place of more costly 2-inch dimension lumber. It provides a smooth surface for floor and counter coverings and in built-ins and furniture can often be incorporated in designs that reduce the amount of framing required.

Particleboard and medium-density fiberboard (MDF) are what the industry calls "mat-formed" products. In layman's language, this simply means that the materials used in manufacturing them are mixed together and placed on wide belts which carry them to presses, where the materials are compressed under heat into dense, rigid panels. The dimensions of these panels are usually 4×8 feet, and the thicknesses are $\frac{1}{4}$, $\frac{3}{8}$, $\frac{5}{8}$, and $\frac{3}{4}$ inch. All composition boards are manufactured this way, of course, and the differences between them are the result of the materials used and the methods of processing the materials.

Particleboard and MDF have wood chips, splinters, and flakes, as well as screened sawdust (what the industry calls "mill wastes"), as their base. There are different bonding agents used, depending on whether the end product is going to be used inside or outside.

Cured sheets of particleboard move from curing stacks to the edger where they will be reduced to precise 4×8 dimensions.

National Particleboard Mfrs. Assn.

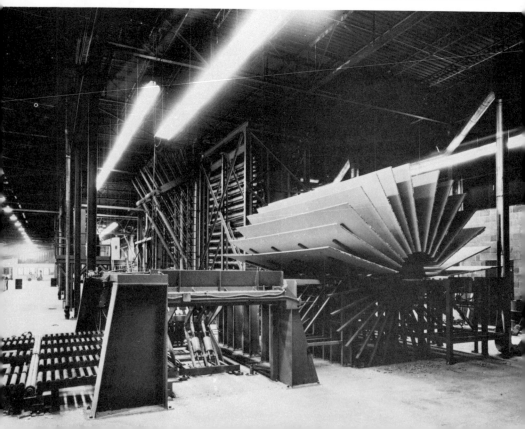

Because its surface shows the wood chips and particles that make up its composition, particleboard is variously called "flakeboard," "splinterboard," and other names.

National Particleboard Mfrs. Assn.

Because of the rather coarse wood chips which go into particleboard, this material is often mistakenly called chipboard, which is really a paper product, a form of cardboard or paperboard. Again, because of the wood flakes embedded in particleboard and visible on its surface, it is often referred to as flakeboard or splinterboard. So, if you've known particleboard only by one of its aliases, you can now discard these in favor of its correct name.

As noted earlier, particleboard can be engineered to special-purpose uses, but this is generally done only for users who can contract for a very large quantity of material. The particleboard you'll find available from your lumber dealer will in all likelihood be one of the two forms which meet the widest range of average uses, or will be a medium-density fiberboard.

As density is the equivalent to grade in lumber and plywood, this is the starting point for establishing the characteristics of particleboard and MDF. There are only two types of these products. Type 1 is made with a urea-formaldehyde resin binder and is for interior use. Type 2 is made with a binder of phenolic resin, which is resistant to both moisture and heat; Type 2 is therefore what you'd choose for exterior use. The specifications for both types are similar, since this is a manufactured rather than a natural or total wood product.

Specifications. Before we get to the chart which will detail the specifications for particleboard and MDF, some translations of the terms it will use may be helpful:

Class is determined by a combination of the product's overall specifications, which follow.

Modulus of rupture is a measure of the strength of the board in terms of breakage across its surface; it is expressed in pounds of load pressure per square inch (psi).

Modulus of elasticity is a measure of resistance to deflection—in other words, the panel's relative rigidity or flexibility.

Internal bond is a measurement of the forces required to pull apart a panel; the strength of the bond created by the resins which bond the solids into a panel.

Linear expansion is the average dimensional change—shrinking or expansion—which can be expected in response to atmospheric or environmental humidity.

Screw holding measures the force in pounds of pull required to yank a #10 metal screw out of the face and edge.

These specifications are stamped on particleboard panels as a three-digit code: 1-A-1, 1-A-2, 1-B-1, etc. The first digit is always a number, and tells you the *type* of board, that is, which of the two adhesive bonds has been used in making it. The letter is the second digit, and it always indicates the density, which is expressed in terms of weight per cubic foot; the heavier weight indicates a greater ratio of adhesive to wood particles, and thus a heavier, denser panel. The final digit, a number, expresses the sum of the factors built into the panel during manufacture: its modulus of rupture, modulus of elasticity, and internal bond. In other words, the final number tells you the actual *strength* of the panel.

Note on the chart (page 82) that there is a definite ratio or relationship between each set of these figures. The lower the internal bond, the greater the flexibility of the board becomes. Flexibility, of course, indicates generally how much distance a material will span safely between its supports. As a general rule, the denser a material is, the more rigid it is, and the greater load it will support without bending or breaking. In natural materials, such as lumber and plywood, rigidity can be obtained only by using thicker lumber or plywood with more plies. In man-made materials such as particleboard, it can be built into the product.

Choosing particleboard for specific jobs. Using the specifications chart, judging the strength of the particleboard in relation to its thickness, will enable you to select the most suitable kind for the job you have in mind. In selecting, remember that particleboard does not always have to be as thick as lumber or plywood in many applications. You should also realize that from the practical standpoint, only the largest retail suppliers will normally carry in stock more than two grades and types of particleboard in a full range of thicknesses. Generally, you'll find 1-B-1, 1-B-2, and 2-B-2 panels in most retailers' stocks. These are the average particleboards suitable for the widest range of applications. If you can't get the grade you planned, choose the next strongest.

Choose 1-B-1 for floor underlayment, countertops, partitions, and general-utility purposes where no moisture is likely to be encountered. Go to 2-B-2 for use in humid or wet surroundings, or for shelving, cabinets, and furniture cores that are to be covered with a veneer or laminate; 2-B-2 combines greater rigidity with lighter weight. Incidentally, covering particleboard with a veneer or laminate, even with a sheathing of such a thin material as contact paper, adds to its rigidity.

For shelving, cabinet tops, or furniture cores, your most satisfactory choice will probably be medium-density fiberboard, which has specifications midway between 1-B-1 and 1-B-2 particleboard. MDF is used increasingly by home craftsmen for shelves, cabinets, and similar applications. The chart on page 84 shows the shelf spans recommended by manufacturers of particleboard and MDF.

To help you use this chart in your figuring, typical shelf loads in pounds per square foot (lbs/sq ft) are: kitchen cabinet, 15; bedroom closet, 25; and in bookcases or record cabinets, 40. Using the chart for $^1/_2$-inch particleboard, you can see that a safe span for a kitchen-cabinet shelf could be as much as 40 inches, while the maximum span for a bookcase would be 24 inches. However, by using $^3/_4$-inch material, a span of 40 inches could be used in a bookcase.

Working with particleboard. To save your power-saw blades, use only a tungsten-carbide-toothed blade when cutting particleboard. The wear is not as pronounced on handsaw blades, which don't generate as much friction in use. In a saber saw, use a coarse metal-cutting blade. Blades for this latter type of saw are inexpensive; you can replace two or three for the cost of having a circular-saw blade resharpened. You'll get cleaner cuts with both types of power saws if you advance the saw slowly. In drilling, if you have a variable-power drill, use it at top rpm—1000 to 1300. A router used in working particleboard should be fitted with a tungsten-carbide bit. Because of the adhesives used in the binding process, particleboard quickly dulls even the best steel blades.

For fine handwork, if required, use a Stanley Surform tool, which has an easily replaced blade. Do the final smoothing with a file, instead of subjecting a plane blade to the wear these boards will give it. You will find that a mill bastard file is better than a rasp in rough-edging; rasps tend to shatter the edges of particleboard, and a smooth-cut file will get clogged too quickly. When sanding particleboard, an open-faced paper will minimize clogging. You can safely use grits as coarse as #80 without shredding. By far the best way to work particleboard is with strokes parallel to the edges.

Butt joints, even if glued and screwed or nailed, are not recommended for cabinets or freestanding pieces of furniture with particleboard cores. Professional cabinetmakers and manufacturers use a variety of special interlocking and shouldered joints, but these aren't practical for the home craftsman with a limited toolchest. There are also a number of patent metal clip-and-screw combinations used in factories, but these have not yet reached the home-crafts marketplace. The simplest of these, a metal right-angle brace, is commonly used by manufacturers who spe-

CHARACTERISTICS OF PARTICLEBOARD

TYPE (USE)	1 — Mat-formed particleboard (generally made with urea-formaldehyde resin binders) suitable for interior applications.								2 — Mat-formed particleboard made with durable and highly moisture- and heat-resistant binders (generally phenolic resins) suitable for interior and certain exterior applications when so labeled.			
DENSITY (GRADE) (min. avg.)	HIGH-DENSITY, 50 LBS/CU FT AND OVER — a		MEDIUM-DENSITY, BETWEEN 37 AND 50 LBS/CU FT — b		LOW-DENSITY, 37 LBS/CU FT AND UNDER — c				HIGH-DENSITY, 50 LBS/CU FT AND OVER — a		MEDIUM-DENSITY, BETWEEN 37 AND 50 LBS/CU FT — b	
CLASS Strength classifications based on properties of panels currently produced.	1	2	1	2	1	2			1	2	1	2
MODULUS OF RUPTURE (min. avg.) psi	2400	3400	1600	2400	800	1400			2400	3400	1800	2500
MODULUS OF ELASTICITY (min. avg.) psi	350,000	350,000	250,000	400,000	150,000	250,000			350,000	500,000	250,000	450,000

82

	INTERNAL BOND (min. avg.) psi	LINEAR EXPANSION (max. avg.) percent	SCREW HOLDING FACE (min. avg.) lbs.	SCREW HOLDING EDGE (min. avg.) lbs
	60	0.25	250	200
	65	0.35	225	160
	400	0.55	500	350
	125	0.55	450	-
	30	0.30	175	-
	20	0.30	125	-
	60	0.30	225	200
	70	0.35	225	160
	140	0.55	-	-
	200	0.55	450	-

MAXIMUM SHELF SPANS IN INCHES FOR UNIFORM LOADING

	END SUPPORTED — MAXIMUM SPAN									MULTIPLE SUPPORTS — MAXIMUM SPAN									OVERHANGING SHELF — MAXIMUM OVERHANG								
	Type 1-B-1			Type 1-B-2			MDF			Type 1-B-1			Type 1-B-2			MDF			Type 1-B-1			Type 1-B-2			MDF		
LOAD°	1/2	5/8	3/4	1/2	5/8	3/4	1/2	5/8	3/4	1/2	5/8	3/4	1/2	5/8	3/4	1/2	5/8	3/4	1/2	5/8	3/4	1/2	5/8	3/4	1/2	5/8	3/4
50.0	14	18	21	17	21	25	15	19	23	19	24	29	22	28	34	20	25	30	6	8	10	8	9	11	7	9	10
45.0	15	18	22	17	21	26	16	19	23	20	25	30	23	29	35	21	26	32	7	8	10	8	10	12	7	9	11
40.0	15	19	23	18	22	27	16	20	24	20	26	31	24	30	36	22	27	33	7	9	10	8	10	12	7	9	11
35.0	16	20	24	19	23	28	17	21	25	21	27	32	25	31	38	23	28	34	7	9	11	8	11	13	8	10	12
30.0	17	21	25	20	24	29	18	22	27	22	28	34	26	33	39	24	30	36	8	10	11	9	11	13	8	10	12
25.0	18	22	26	21	26	31	19	23	28	24	30	36	28	35	42	25	32	38	8	10	12	9	12	14	9	11	13
20.0	19	24	28	22	28	33	20	25	30	26	32	38	30	37	45	27	34	40	8	11	13	10	13	15	9	12	14
17.5	20	25	29	23	29	34	21	26	31	27	33	40	31	39	46	28	35	42	9	11	14	11	13	16	10	12	14
15.0	21	26	31	24	30	36	22	27	33	28	35	41	33	41	48	30	37	44	9	12	14	11	14	17	10	13	15
12.5	22	27	32	26	32	38	23	29	34	29	37	44	35	43	51	31	39	46	10	12	14	12	15	18	11	13	16
10.0	23	29	34	27	34	40	25	31	37	31	39	46	37	46	54	33	41	49	11	13	16	13	16	19	11	14	17
7.5	25	31	37	30	37	43	27	33	39	34	42	50	39	49	58	36	45	53	12	14	17	14	17	20	12	15	18
5.0	28	34	41	33	40	48	30	37	43	38	46	55	44	54	64	40	49	58	13	16	19	15	19	22	14	17	20

SHELF THICKNESS IN INCHES

°Load in pounds per square foot

cialize in knocked-down pieces that are shipped flat for home assembly. These differ from the thick narrow metal angles used as braces on wooden furniture in that they have a wide, flat bearing surface and are very thin. Even these have not yet appeared on the home-crafts market, to my knowledge.

By far the easiest and most satisfactory corner joint for the home craftsman to use is a simple rabbet joint or a tenon joint. Rabbet joints should be glued and held with finish nails while the glue dries. The nails are left in place, of course; their heads are countersunk and the holes are filled before a covering is put on. The illustration shows this joint, which can be cut with a router or even with a circular saw having a tungsten-carbide blade.

About as easy to cut, and perhaps a bit stouter, is a tenon joint, also illustrated. This joint too can be cut with a few passes of a circular-saw blade. The joint should offset, as illustrated, and glued. Clamps should be used while the glue sets, though with the new fast-drying wood glues, drying time is a matter of minutes instead of hours.

Installing particleboard. When using particleboard for floor underlayments, built-in cabinets, seats such as window seats or dinette benches, bookcases, or similar permanent fixtures, the most satisfactory method to use is a combination of gluing and nailing. With glue, fewer nails are needed and there is no tendency for a joint to pull loose as sometimes happens when nails alone are used.

Floor underlayment over which carpet or vinyl covering is to be laid should be fitted so that at no point do four corners of the panels come together. This is done by cutting the first panel in half to make a 4×4-foot section. Butt the next panel in line to the half-section, and start the second course with a full 4×8-foot panel parallel to the first course. This should be done whether nails alone or glue and nails are used.

A hard-setting casein glue is recommended, applied with a roller where the panels butt together. Before laying the particleboard, the subfloor should be cleaned well. After installation, the glue joints should be nailed on about 16-inch centers to ensure good contact. When installing a cabinet or other built-in, the back of the piece can be glued to the studs and nails on 10-inch to 12-inch centers driven to maintain contact while the glue sets.

Shelves in built-in cabinets should be installed in dado cuts during construction of the cabinet, and glued. Usually, no nailing is required in this type of installation. In an earlier section of this chapter, the span of shelves was charted. Keep in mind that particleboard shelves can generally be of thinner stock than lumber or plywood shelves.

Finishing particleboard. Although laminate sheeting, which will be covered in a later chapter, is usually applied over particleboard, there are many other options available. Contact paper is one of these. So is

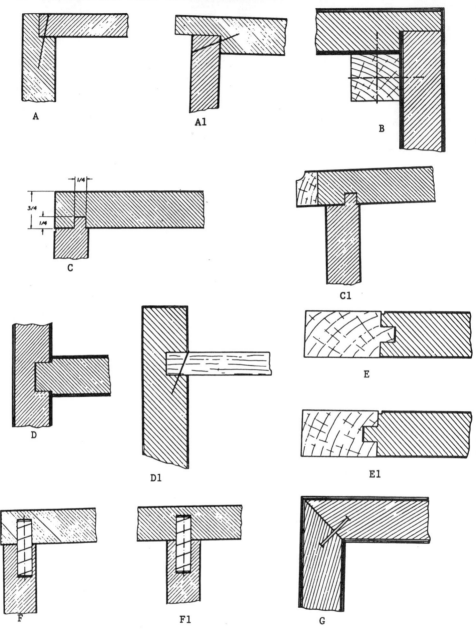

American Hardboard Assn.

Joints recommended for use in constructing cabinets, furniture, pieces using particleboard as a core material are shown above. Figures *A* and *A1* are nailed rabbet joints; *B*, a reinforced rabbet joint; *C* and *C1*, tenon joints, which should be cut as shown, with the tenon offset; *D* and *D1*, dado joints used in shelves; *E* and *E1* are dado joints used in edge-finishing; *F* and *F1*, dowel joints; *G*, miter joint reinforced with staple, spline, or spline-nail. All joints should be glued.

Four key steps in installing particleboard underlayment for a floor covering of vinyl roll or tile.

1. Nail along edges about 8″ OC, using ring-shank or screw-shank nails.

2. Apply a hard-setting casein glue along joints, using a roller.

3. Sand glued joints smooth.

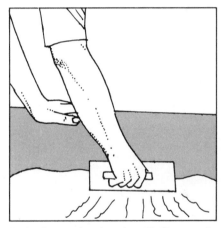

4. Apply mastic, then install tile or roll flooring.

stain, and so is paint or varnish. However, water-based paints are not recommended. Any particleboard surface covered with self-adhering contact paper should be sanded smooth, wiped free of dust, and given a sealer coat of shellac to assure good adhesion.

If a very smooth glossy surface is required, particleboard should be sanded and filled with a sanding sealer before putting on the undercoat. Most particleboards have a small quantity of paraffin wax added during manufacture as a moisture retardant, so if a paint containing solvents which are also wax solvents is used, some of this wax may be absorbed by the paint film, and slow drying will result. A shellac sealer coat will prevent this, as shellac is not a wax solvent. Do not use a shellac barrier coat on exterior surfaces, though, for it will reduce moisture resistance.

Particleboard to be painted should have an undercoat and a finish coat for best results, though if you want a textured effect the undercoat can be omitted. A water-based paint will usually cause a bit of surface rising, if a more pronounced textured effect is wanted. You can paint in any style that suits you best, with brush, roller, or spray gun.

Staining particleboard is commonplace and easy, but use an oil-based rather than a water-based stain to achieve a smooth surface. Varnish may be applied over stain, or over the board itself for decorative effect or protection against moisture.

Edges of particleboard will absorb paint or stain unevenly. They should be filled with a stainable wood paste and sanded, or covered with a wooden trim strip or a glued veneer trim strip. Of course, edges of countertops and similar built-ins are usually covered with a metal strip.

For exterior painting of exposed particleboard, use an oil-based enamel undercoat or outside white primer to give surface protection and provide a smooth base for the finish coats.

One important thing to keep in mind, though, is that particleboard has a much rougher surface before sanding than do most other wood surfaces such as lumber and plywood. If left unfinished, an exposed particleboard surface soon gets dirty, and is very hard to clean. For this reason, if no other, all particleboard surfaces that are exposed, or subject to use, such as cabinet interiors, need to be finished in some manner.

HARDBOARDS. Hardboards are manufactured panels, made by one of two processes, wet or dry. Both kinds begin with the same basic material: wood chips of reasonably uniform size. The chips are about the size of a man's thumbnail, and may be produced by chipping logs or by using odd-sized pieces left over when logs are sawed into boards. Some sawdust may also be added. In one of the two processes, the chips are reduced to wood fiber by steaming and grinding; in the other, the fibers are formed by loading the chips into a high-pressure cylinder into which steam is piped and suddenly released in an explosion which reduces the chips to fibers.

These fibers are formed into mats by either a wet or a dry process. In the former, the fibers are aligned on a mesh screen over which water flows, and then drained and squeezed dry. In the second, the dry process, air is used instead of water to arrange and align the fibers. The end result of both these processes produces virtually identical hardboard panels.

After aligning, the wood fibers are bonded with lignin, which is an adhesive produced from wood. The resulting pastelike mat is subjected to heat and pressure which in effect weld the fibers into a dense, solid, smooth panel. In these two final manufacturing steps, other materials are often added to produce differences in density, reduce or increase rigidity, or to alter surface properties.

When the panels emerge from the hot presses they contain almost no moisture, and must be seasoned to approximate atmospheric humidity. This reduces warping or buckling. The panels are trimmed to standard sizes, 4×7, 4×8, 4×9, 4×10, and 4×12 feet. The 4×8-foot panel is the size most often stocked by local retailers, but the other standard sizes are always available on special order. Standard thicknesses range from $^1/_{12}$ inch, which are the decorative hardboards used for interior finishing, through 1 $^1/_8$ inches, used in industrial applications. Most retailers carry thicknesses of $^1/_8$, $^1/_4$, and $^1/_2$ inch in tempered hardboard and $^1/_4$ and $^3/_8$ inch in pegboard.

Choosing hardboards for specific jobs. Hardboards have an extremely wide range of use in building, interior remodeling, and craftwork. They are produced in five different classes, which in broad terms equal the grading or category systems used in identifying all panel-type products. The chart on page 92-93 will give you a complete listing of the classes, thicknesses, and physical properties of hardboard available and will help you to make the proper choices when you look for hardboards to be used for general or special applications.

Definitions of the "class" designations are those adopted by the hardboard industry and are standard for all manufacturers.

Tempered hardboards are those impregnated with moisture inhibitors, and in which additives have been placed before the bonding process to improve stiffness, hardness of surface, or resistance to abrasion. Generally, tempered hardboards are a bit tougher than those which have not been altered by the additives.

Standard hardboards are those which have not been tempered. These are less moisture-resistant than the tempered boards.

Service-tempered hardboards have undergone essentially the same processes as have the tempered boards, but contain a lower percentage of the additives in the tempering process.

Service hardboards are those without the additives used in tempering. See the chart for the differences in tensile strength and resistance to moisture.

Industrialite hardboards are of medium density, lighter than any of the other hardboards.

Surface tells you whether or not both sides of the panel have been finished. Those described as S1S (smooth one side) will have mesh marks on one side, or one side may show a rough texture caused by abrasion of the press platen during the compression step of its manufacture. S2S (smooth two sides) panels are smooth on both surfaces.

Water resistance is the percentage of swelling that can be expected under conditions of extreme exposure; this test is made by submerging panels in water for twenty-four hours.

Modulus of rupture is the average number of pounds per square inch of pressure required to cause the panel to break across the face.

Tensile strength is the force in pounds per square inch of pull required to break the internal bonding of a panel.

By checking your job-application requirements against the figures given on the chart, you can very quickly determine which class of hardboard will be adequate for your needs. There is no point in paying extra for increased moisture resistance, for example, if the hardboard is being applied in any room except the bathroom inside the house or in a storage area. Using the chart's specifications saves your time in ordering and your pocketbook when paying.

In addition to knowing the characteristics of the hardboard you will be using, it will also be helpful to know the industry terms, which are also the retailers' terms, describing various surface treatments available. Beyond the mere fact that a panel is smooth on one or both sides—S1S or S2S—there are a number of factory finishes or surfaces that you can buy to make your finishing job easier or to give you a decorative panel or partition.

Perforated is the correct designation for pegboard, with which every home craftsman is familiar.

Filigree describes hardboard perforated in ornamental designs or patterns; these panels are used as room dividers, in entry halls, as outdoor vestibules, and so on.

Patterned hardboard has a textured surface pattern that has been pressed or machined. It is used both in exterior siding and as a wainscot or a room divider where a decorative motif is desired.

Embossed hardboard has a pressed-in pattern, such as simulated leather graining, basketweave, woodgrain, etc.

Grooved hardboard is used chiefly as an exterior siding; it can be applied with the grooves vertical or horizontal.

Striated is the same pattern of shallow random-spaced grooves found in plywoods that bear this designation.

Primed or *coated* hardboard has been given a base coat of paint or a coating which makes painting easier.

Factory-sealed or *factory-filled* describes a hardboard which has been treated during manufacture to provide an easily painted surface.

Not all retailers will keep every type of hardboard in stock, but each of the classes and surface treatments which have been described here are available on special order.

Working with hardboard. While hardboard is similar to particleboard in that both have a bonding agent which holds their solid materials together, the lignins used in bonding hardboard are not as hard on edged tools as are those in particleboard. A tungsten-carbide saw blade is not necessary when working hardboards with power tools. You should, however, use a fine-cut blade rather than a combination blade in order to avoid fuzzing the edges of the saw cut. If a jigsaw is used, a fresh metal-cutting blade will give you a virtually fuzz-free cut; so will a fine-tooth wood blade. If a handsaw is used, it should be a fine-toothed crosscut saw.

When sawing any surface-finished hardboard, the face should be up when a handsaw is used, and down if a portable power saw is used. By observing this procedure, almost no fuzz will be left on the finished side. When hand-sawing, apply pressure to the blade only on the downstroke.

For drilling hardboard, a hand brace and auger bit is the most satisfactory tool. Next in line come speed bits in a power drill. Start the hole on the face that will be exposed, and use a block of scrap wood to back up the area being drilled; this will minimize burred edges.

Hardboards that must be edge-planed are most easily worked with a block plane rather than a bench plane or jack plane. However, a file will reduce any saw marks or irregularities almost as quickly as a plane. So will coarse sandpaper used with a block or in a power sander. Hardboards do not clog sandpaper, so there is very little difference between using open-faced or closed-faced sandpaper. Auto-body finish paper, #400 wet-or-dry, used wet, will produce a high polish on hardboard and at the same time prepare the surface for painting.

Always use casing nails rather than finishing nails when nailing hardboard where nailholes must be filled. You'll find that the cone-shaped heads of casing nails push up a smaller burr around the holes than do the more abruptly shouldered heads of finishing nails. Always predrill for both nails and screws; hardboard is exceptionally dense, and predrilling will save a great deal of time in nailing.

Applying hardboard. Because decorative-surfaced hardboard has been classed as a panelboard for the purpose of this book, its application will be covered in the following chapter. Here, we'll go into the application of hardboards used as floor underlayment and exterior siding, and we'll also mention pegboard.

Because it is slightly flexible when used over a very irregular subfloor as an underlayment for vinyl tiles or roll flooring, hardboard used for this purpose should be relatively thick; an S1S Service or Industrialite type $^7/_{16}$ or $^1/_2$ inch thick is indicated. Over a subfloor that is not warped, a thinner panel can be used—as thin as $^1/_4$ inch if the subfloor is quite smooth.

Nailing, using ring-shank flooring nails, is suggested when using hardboard as an underlayment over a rough or warped floor. On a smooth floor, thin underlayment can be installed very quickly using vinyl flooring adhesive. This will save the minor nuisance of using two different adhesives. The adhesive can be applied with a roller fitted with a long handle. Apply one course to the subfloor along the lines where the seams

CLASSIFICATION OF HARDBOARD BY SURFACE FINISH, THICKNESS, AND PHYSICAL PROPERTIES

CLASS	SURFACE	NOMINAL THICKNESS (inch)	WATER RESISTANCE (MAX AV PER PANEL) — WATER ABSORPTION BASED ON WEIGHT S1S (percent)	S2S (percent)	THICKNESS SWELLING S1S (percent)	S2S (percent)	MODULUS OF RUPTURE (MIN AV PER PANEL) (psi)	TENSILE STRENGTH (MIN AV PER PANEL) PARALLEL TO SURFACE (psi)	PERPENDICULAR TO SURFACE (psi)
1 Tempered	S1S	$1/12$	30	—	25	—	7000	3500	150
	S1S and S2S	$1/10$	20	25	16	20			
		$1/8$	15	20	11	16			
		$3/16$	12	18	10	15			
		$1/4$	10	12	8	11			
		$5/16$	8	11	8	10			
		$3/8$	8	10	8	9			
2 Standard	S1S and S2S	$1/12$	40	40	30	30	5000	2500	100
		$1/10$	25	30	22	25			
		$1/8$	20	25	16	18			
		$3/16$	18	25	14	18			
		$1/4$	16	20	12	14			
		$5/16$	14	15	10	12			
		$3/8$	12	12	10	10			
3 Service-tempered	S1S and S2S	$1/8$	20	25	15	22	4500	2000	100
		$3/16$	18	20	13	18			
		$1/4$	15	20	13	14			
		$3/8$	14	18	11	14			

THICKNESS TOLERANCES FOR HARDBOARD PANELS

Class	Surfaces	Nominal thickness							
4 Service	S1S and S2S	1/8	30	30	25	25	3000	1500	75
		3/16	25	27	15	22			
		1/4	25	27	15	22			
		3/8	25	27	15	22			
		7/16	25	27	15	22			
		1/2	25	18	15	14			
	S2S	5/8	–	15	–	12			
		11/16	–	15	–	12			
		3/4	–	12	–	9			
		13/16	–	12	–	9			
		7/8	–	12	–	9			
		1	–	12	–	9			
		1-1/8	–	12	–	9			
5 Industrialite	S1S and S2S	3/8	25	25	20	20	2000	1000	35
		7/16	25	25	20	20			
		1/2	25	25	20	20			
	S2S	5/8	–	22	–	18			
		11/16	–	22	–	18			
		3/4	–	20	–	16			
		13/16	–	20	–	16			
		7/8	–	20	–	16			
		1	–	20	–	16			
		1-1/8	–	20	–	16			

NOMINAL THICKNESS	THICKNESS TOLERANCE (MIN.-MAX.)
inch	*inch*
1/12 (0.083)	0.070-0.090
1/10 (.100)	.091- .110
1/8 (.125)	.115- .155
3/16 (.188)	.165- .205
1/4 (.250)	.210- .265
5/16 (.312)	.290- .335
3/8 (.375)	.350- .400
7/16 (.438)	.410- .460

NOMINAL THICKNESS	THICKNESS TOLERANCE (MIN.-MAX.)
inch	*inch*
1/2 (.500)	.475- .525
5/8 (.625)	.600- .650
11/16 (.688)	.660- .710
3/4 (.750)	.725- .775
13/16 (.812)	.785- .835
7/8 (.875)	.850- .900
1 (1.000)	.975-1.025
1-1/8 (1.125)	1.115-1.155

will fall. The hardboard should be edge-nailed on 16-inch to 18-inch spacing.

Arrange the layout of the panels so that there are no spots where four corners come together. Do this by cutting the first sheet laid in half to make a 4×4-foot square. This will automatically ensure that the succeeding courses will have no more than two contiguous corners. See the diagram in Chapter 3 for details.

When a pegboard wall is being installed, cut small squares from waste trimmed in fitting the board and use them for spacers to keep the center of the board rigid. Predrill holes in the pegboard and glue the squares, using two in each place you need a spacer, to the back of the board, centered on the holes. Base-shoe, quarter-round, or cover molding can be used to frame the edges of the pegboard and avoid an unsightly gap between it and the wall on which it's mounted.

Hardboard siding can be installed on studding of new buildings or over old siding when remodeling. In new construction, check local building codes for sheathing and vapor-barrier requirements. When remodeling, a vapor barrier costs little and not only may make your home more comfortable, but may prolong the life of its basic structure. You can install hardboard siding over masonry walls by using furring strips, and in the case of hardboard—or any siding, for that matter—applied over masonry, a vapor barrier is an excellent idea.

There are two basic types of hardboard siding: panels and lap siding. The general technique of applying panel siding has been covered in Chapter 3, and that used in applying hardboard siding differs very little. The chief difference is that when hardboard panels are used, batten joint strips are recommended. The details of installing these is shown in the accompanying sketches.

Square-edge panel siding should be nailed in either vertical or horizontal courses to studding in new work and to battens when remodeling. Old walls may not be true, and shims may be necessary to ensure straight courses. So may a new foundation-line starter strip. A $1/16$-inch gap should be left between panels and caulked. A sketch also shows the recommended method of applying shiplapped siding panels.

Lap-siding installation is also covered in the accompanying sketches and photos. It should be applied on studding spaced on 16-inch centers; on existing walls or battens in remodeling, the same spacing should be used. Joint moldings of thin aluminum are recommended when lap siding is installed; again, refer to the illustrations for application details, including such matters as spacing of joints, placing of starter strip, and nailing.

Corners of lap siding can be finished several ways. There are metal strips for both inside and outside corners, or inside joints can be butted to a corner board and outside corners finished off with butted battens. The illustrations show both ways.

INSTALLING HARDBOARD PANEL SIDING

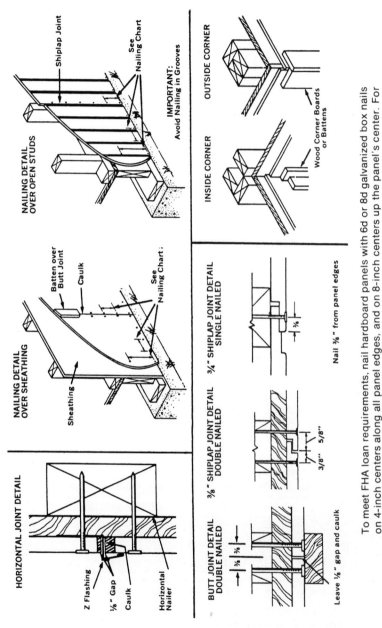

NAILING DETAIL OVER OPEN STUDS

Shiplap Joint

See Nailing Chart

IMPORTANT:
Avoid Nailing in Grooves

NAILING DETAIL OVER SHEATHING

Batten over Butt Joint

Caulk

See Nailing Chart

Sheathing

HORIZONTAL JOINT DETAIL

Z Flashing

⅛" Gap

Caulk

Horizontal Nailer

BUTT JOINT DETAIL DOUBLE NAILED

⅜" ⅜"

Leave ⅛" gap and caulk

⅜" SHIPLAP JOINT DETAIL DOUBLE NAILED

3/8" 5/8"

¾" SHIPLAP JOINT DETAIL SINGLE NAILED

⅜"

Nail ⅜" from panel edges

INSIDE CORNER

OUTSIDE CORNER

Wood Corner Boards or Battens

To meet FHA loan requirements, nail hardboard panels with 6d or 8d galvanized box nails on 4-inch centers along all panel edges, and on 8-inch centers up the panel's center. For all practical purposes, structural integrity will be assured with 6d or 8d galvanized box nails on 6-inch centers along edges and on 12-inch centers in center of panel. When panels are applied horizontally, flashing should be used to supplement caulking as shown in the sketch, and cross-members should be installed to accept nails at edges and along center lines of panels.

American Hardboard Assn.

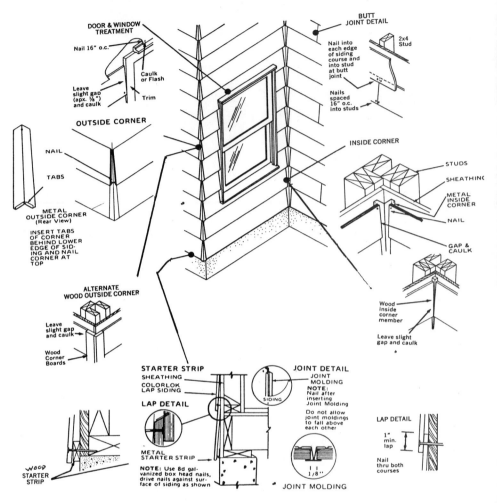

American Hardboard Assn.

By referring to the sketches above you can determine not only the method of applying hardboard lap siding, but you can see your options in handling joints and corners as well as other details. Nailing details are also shown. The series of photos that follows will be helpful in showing how some of the details are handled.

Because of its great rigidity, time and money can be saved by using exterior hardboard panels when putting up utility buildings. Usually, studding for these small structures can be spaced 24 inches on center rather than the usual 16 inches. Such small buildings are usually exempt from building codes, and you sacrifice neither safety nor service by using the more economical stud spacing, which results in great structural integrity.

Finishing hardboard. A successful finish on hardboard—unless you're using one of the factory-primed or sealed types—depends on pre-

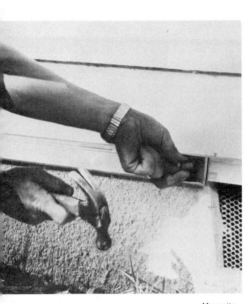

Metal starter strips can be used either over existing siding or over a wooden starter strip; these give the key first course extra stability and minimize edge-curl caused by dampness.

Leave a $1/_x$-inch gap between boards when using metal joint strips. Slide the strips into the gap, nail at top of strip. Later, a thin bead of caulking should be put along each edge of the metal strip.

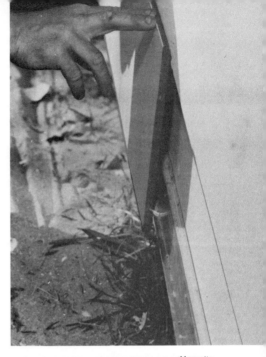

When metal starter strips are used, no bottom nailing is required. The nailing should begin at a corner, as shown.

Slide metal corner strips in from bottom, and make sure bottom flanges engage the corners of the boards. Later, put a thin bead of caulking along the edges of the metal.

Masonite

Install J-flashing over windows and doors, caulk the edges between frame and old siding or sheathing before nailing the metal J-strip in place. Trim siding to fit, nail at both top and bottom along window or door frames.

application surface preparation. Unless you take the time to give the surface "tooth" by sanding, the impervious board is the very devil to paint. Use a wet-or-dry auto-body paper, grade #380 or #400, with a sanding block. Use the paper wet. Keep a bucket of water handy and dip the paper every few minutes to remove clogging from its grits. You don't need to apply a lot of pressure; use just enough to break down the natural sheen that is characteristic of the surface of hardboards.

Cover with an undercoat designed for use with the kind of paint that will be used as a finish coat. If enamel is going to be used as a finish coat—and about 90% of the time it will be—sand the undercoat. You'll probably need two coats of enamel to get a slick, smooth job.

Any kind of laminate or roll vinyl material can be used in surfacing countertops, desktops, etc. made from hardboard. Use contact cement with laminate, vinyl adhesive with the roll material. Both should be rolled for a thorough bond; a rolling pin will do the job adequately, or the surface can be tapped with a padded block to secure good adhesion. There is a variety of edge moldings of both wood and metal that can be used to finish off the edges. For full details of laminate installation, see Chapter 6.

SOFTBOARDS. In recent years the softboards have become the stepchildren of building materials. These are panels made from finely ground wood pulp, held together with a bonding agent and lightly compressed. Their density is far less than that of either particleboard or hardboard; except for the laminated exterior sheathing which falls into this family, their structural strength is very slight.

Although at one time softboards were a major item in the stock of every retail lumber outlet, their jobs have for the most part been taken over by newer and more durable materials. Today's most common use for softboards is for sound control as ceiling panels or ceiling tiles, and for a very effective sheathing as laminated panels that include a vapor barrier.

Probably the widest use of softboards today is as ceiling tile or small panels, usually 18×24 inches. The old-style full 4×8-foot acoustical panel has just about vanished; they are very hard to handle because of their easily marred surface. Today's acoustical tile is not the plain-Jane model with a boringly repetitive pattern of evenly spaced holes. The modern acoustical tile and panel conceals its utilitarian function with patterned swirls and ridges and pebbled textures that give ceilings a finish which requires no overhead painting.

There are several types of lattice suspension systems for the installation of these textured panels. Each one is slightly different, which makes it impractical to go into detailed installation techniques here. These systems all share a common feature; they are all methods of fitting small

Celotex

Above: Typical patterns of softboard available in ceiling tiles for installation with staples.
Below: Sample patterns in 2×4 panels used in suspended ceilings.

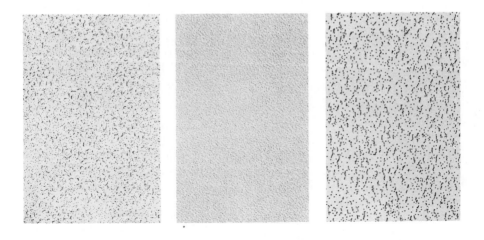

First step in putting up a softboard tile ceiling is the installation of furring strips. These should be applied whether the job is being done over new ceiling joists as shown, or over an existing ceiling. The strips should be spaced 12 inches on center and run the length of the room rather than the width.

softboard panels, usually the 18×24-inch size mentioned earlier, into an interlocked latticework of T-shaped center supports and L-shaped wall supports. The supporting framework is suspended at intervals by metal straps or holders that are attached to the rafters.

Softboard tiles can be stapled directly to existing ceilings that have cracked or warped or simply aged beyond the point of easy renovation. These tiles have tongue-and-groove edges which interlock on two sides, and their installation with a stapling gun goes very quickly. On ceilings that have warped or are badly cracked, furring strips of lath may be necessary to give a level installation. A variety of pressed, carved, and textured surfaces are available, and the tiles are prefinished, usually in white.

Exterior softboard. This type of softboard is a laminated product with exterior coatings of asphaltic compounds. They may or may not have one or more center plies of the same material. Two types are available, one suitable for sheathing that combines a vapor barrier, the other suitable for use as a combination sheathing/siding where local building codes permit this type of construction. These sheets are widely used in rural areas for farm utility buildings. Exterior finishes are prefinished; the most readily available are a pebbled surface and a grooved surface that simulates brick.

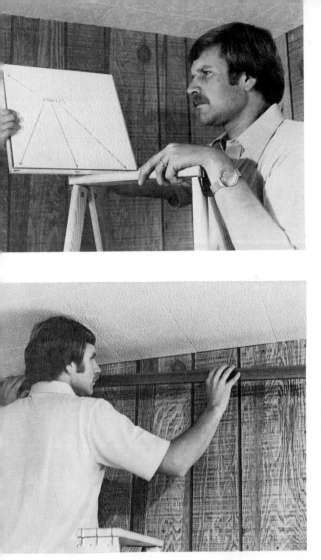

Staple the tiles to the furring strips in the pattern shown. The tongue-and-groove joint on the sides opposite the stapled edges will provide all the support the tiles require.

Celotex

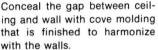

Conceal the gap between ceiling and wall with cove molding that is finished to harmonize with the walls.

Celotex

Panels are available in 4×8-foot, 4×10-foot, and 4×12-foot sizes, $^5/_8$ inch and $^3/_4$ inch thick. They are nailed to studs that should be no more widely spaced than 16 inches on center, for these softboard panels have relatively little structural strength compared, say, to hardboard or particleboard. When installed as siding, edges should be butted with a gap of about $^1/_8$ inch and the void filled with grout, then covered with a wood batten. Corners should be butted and outside corners finished with a metal molding or corner boards; inside corners should also be finished with a square corner board or a metal molding.

Although these softboard panels are most useful as sheathing because they include a vapor barrier and have good insulating qualities, the siding version provides an inexpensive and fast way to erect a temporary building or a utility building.

101

INSTALLING A SUSPENDED CEILING

1. With a chalkline and spirit level, establish a line around the room approximately 3 inches below the level of studs or an existing ceiling. If light fixtures are to be recessed in the new ceiling, allow 6 inches.

2. Install the right-angle ceiling strip by nailing. Take pains to get the strip level.

3. With a chalkline establish parallel lines 24 inches apart on joists or old ceiling if using 2×2 tile, 48 inches apart if using 2×4 panels.

4. Drive nails (or more expensive screweyes) every 4 feet along the center chalkline and make lattice hangers of #14 gauge wire as shown. When the wire is in place, clench the nailheads into the joists as shown in the next picture.

5. Install lattice strips as illustrated on the wires. Use a spirit level to get the strips straight; bend the wires as required to level them.

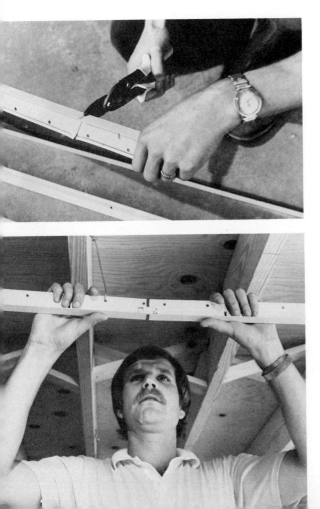

6. Use tinsnips as shown to cut the lattice strips where necessary; cut the thin top section, then the exposed angle strip. Make one cut on each side.

7. End splice tabs are provided to make tight joints in the lattice strips; use these according to manufacturer's directions.

Courtesy of Celotex

8. Space the initial course of lattice strips so that the end connectors of the ready-cut cross-strips engage in the preformed slots.

9. When the latticework is completed, put in the panels by inserting them at an angle.

10. Cut openings for pipes, light fixtures, or other objects that must pass through a ceiling with keyhole saw or craft knife.

5 | Choosing and Using Gypsum Board and Prefinished Panels

BASICALLY, GYPSUM BOARD is a sandwich. Gypsum is a mineral, and in mineral terms its proper name is hydrous calcium sulfate, which means that it mixes readily with water or other liquids. In its most familiar form, as plaster of paris, gypsum is used to make castings of statuary, vases, molds for patterns, and splints for broken bones. When it dries, gypsum is quite brittle, particularly when its dry form is thin, as in gypsum sheets. It is also quite soluble in water in its unaltered state, but when water-resisting chemicals are mixed with it, gypsum becomes relatively resistant to anything except prolonged exposure to dampness.

Gypsum board is made by bonding sheets of gypsum between two thin cover sheets of cardboard. The gypsum is treated with water-repelling agents, and the cardboard side which will face into the room is processed to accept paint smoothly.

Manufacturers of gypsum board are reluctant to go into exact details of their production processes, or to allow publication of pictures of their operations. However, a fair deduction can be made that the process involves a continuous extrusion of the gypsum mix through wide dies which form the material into a ribbon 4 feet wide. The extruded ribbon is deposited on a cardboard strip, and a second strip is laid on top. The ribbon then passes through a drying press, in which the cardboard is bonded to the gypsum ribbon while excess moisture is being baked from the still-soft core. The tapered edges drywall boards have are very probably formed by the die, and a final step is to close the long edges of the panel before it is cut into market lengths.

This basic process is, of course, subject to variations, as a number of different types of gypsum boards, engineered for both interior and exterior use, are manufactured. Drywall panels are also available factory-finished on one side, which would involve the additional step of applying the decorative facings.

EXTERIOR GYPSUM BOARD. While gypsum boards are primarily associated in the minds of most with drywall interior construction, there are two types of these panels made for exterior use as sheathing, and a third type designed to be used for exterior ceilings such as in porches, carports, soffits, and the open-air arcades of shopping malls. All three of these types are weather-resistant; a mixture of asphaltic materials is added to the core and special coverings of moisture-resistant cardboard or paper are used. The paper covers the long edges of the panelboards, and the short edges are treated with a moisture-resistant compound. The ceiling panels are designed to be finished with paint or plaster; the sheathing is, of course, covered by siding, which may be of any kind.

Gypsum-board exterior ceiling panels are manufactured in thicknesses of $1/2$ and $5/8$ inch, and in lengths of 8, 9, 10, and 12 feet. All are 48 inches wide, the industry standard. Like other drywall panels, one side is covered with a paper that is designed to take paint smoothly. The panels can also be covered with texture paint for a stuccolike effect.

Gypsum-board sheathing panels are of two types: triple-sheathed and regular. Each is designed to be nailed to studding spaced 16 inches on center, or to be attached by sheet-metal screws to metal studs. Each type can be used under any kind of siding: masonry, wood, metal, or composition. The triple-sheathed panels are extremely light, weighing only a bit more than $1 1/2$ pounds per square foot, and are $4/10$ inch thick. Thicknesses of regular gypsum board panels is $1/2$ and $5/8$ inch. Both types come in 8-foot and 9-foot lengths 48 inches wide.

DRYWALL GYPSUM BOARD. Almost everyone is familiar with drywall panels, which in their original form did so much to revolutionize interior finishing of homes and other buildings. For many years, the standard method of finishing interior walls was by nailing wooden laths about 1 inch to $1 1/4$ inches wide and spaced approximately $1/2$ inch apart to interior studding, then applying plaster over these horizontal laths. The method was costly in terms of both time and materials. Carpenters finished off a room with the lath, and plasterers came in to apply first a "scratch coat" of plaster over the laths, then a smooth finish coat. After the plaster had dried, the carpenters returned to put on the final touches of baseboards and to install window and door frames. Then the painters and paperhangers finished off the decorating job.

In the late 1930s, the first gypsum boards appeared, though not in their present form. The original gypsum boards were perforated with $3/4$-inch holes, and looked much like king-sized pegboard. Known originally as buttonboards, these first gypsum panels had been developed to replace the increasingly scarce wood that went into laths, and to reduce the increasingly high cost of preparing a room for plastering. The purpose of the holes in the buttonboard was to anchor the plaster scratch coat.

Development of buttonboard into today's gypsum board followed quickly, spurred by the need to conserve materials and manpower during World War II. Smooth-surfaced panelboards needed no double coat of plaster; the board itself provided a suitable surface for either wallpaper or paint. It would be difficult to find a house of less than mansion stature built after the late 1940s which stayed with the by then outmoded interior finish of paint or wallpaper over lath-supported plaster.

Logically, the next step was to do away with the need for painting, and this came with the development of prefinished gypsum-board panels on which a decorative face was factory-applied. We'll look at the methods of installing unfinished drywall paneling in this section, because "standard" drywall has seniority.

Standard drywall material is the cardboard-covered, unpainted gypsum panelboard already described. It is applied to studding with rust-resistant nails: cement-coated, galvanized, or blued, with preference in that order. Panels are made in $\frac{1}{4}$-inch, $\frac{3}{8}$-inch, $\frac{1}{2}$-inch, and $\frac{5}{8}$-inch thickness. Lengths are 8, 9, 10, 12, and 14 feet, and the standard width is 4 feet. The normally used panel is the familiar 4×8-foot size. For installation where moisture is a problem, and for use in remodeling old houses which may not have included a moisture barrier in their original construction, gypsum boards with an aluminum-foil backing are available.

Traditionally, gypsum boards have been applied vertically, but a growing practice is to use the longer panels and install them horizontally. The difference in joint length is substantial; as much as one-third less footage of joints must be nailed when 10-foot, 12-foot, or 14-foot panels are installed horizontally than are required when installing 8-foot panels vertically. Whether the vertical or horizontal method of application is chosen, the method of application is the same.

Installing gypsum board. Few tools are needed to install either the ceiling panels or the wall panels, and the methods used are identical.

Your tool list should include a keyhole saw, a sharp knife, a hammer, a spirit level, and the two special joint knives which look like very wide putty knives. You will also need a straightedge and a measuring tape or rule. Some kind of worktable or a pair of sawhorses will be required to rest panels on while cutouts are being made.

You can saw panels to length if you wish, using the keyhole saw or a power saw, but the easiest way to shorten a panel is by scoring it on the side that will show and snapping off the unwanted portion. Cuts with a

Gypsum board having one face covered with aluminum foil, shown here being installed horizontally, provides an interior wall with a vapor barrier.

U.S. Gypsum Co.

saw must be made around window and door openings and in making cutouts for electrical outlet and switch boxes. These should be made slightly oversize to allow the flanges at the ends or sides of these boxes to clear the cutout.

When scoring a panel to be snapped, use a straightedge. With the panel supported on sawhorses, using a board just short of the scoreline if necessary to span the full width of the panel, score the line with the knife. Use a single stroke if possible. Press the panel firmly in place and snap off the portion that is to be removed with one quick downward push. It will often be necessary to cut the back cardboard free along the snap line.

All joints should fall in the center of the studs. When the panel is placed and its alignment has been adjusted with the spirit level, nail along the tapered edge about $1/2$ inch from the edge. The nails should be spaced 8 to 10 inches apart. Drive a few nails—two or three, widely spaced—into each of the studs under the central portion of the panel.

When nailing, give each driven nail a final sharp tap that will sink it below the surface of the board. This will create a dent or dimple, made by the hammerhead. It will be filled later with joint compound.

When installing ceiling panels, make a pair of T-braces with which to raise the panels and hold them in position while nailing. The braces can be made from 2×4s with a crosspiece of 1×4. Nail the crosspiece either to the end or to the wide face of the 2×4. If nailed to the end, countersink the nails to keep them from marring the board's surface. If the crosspiece is nailed to the side of the 2×4, its edge should protrude a bit above the end so that only a smooth surface is pressed to the board.

Installing gypsum board on ceilings is really a two-person job. Use a brace about 18 inches from each end, raise the panelboard, and position it against the rafters. Wedge the 2×4s in place. If they are about 1 inch longer than the height of the ceiling, a kick on the bottom of the 2×4 will do the job. If the floor is finished and must not be marred, use 2×4s about 4 to 6 inches shorter than the height of the ceiling and secure them with a wedge made by slanting a piece of scrap lumber under the slanted board to hold it firmly in place. The illustrations show how this entire procedure is carried out.

After the panels have been nailed in place, the joints must be filled with joint compound, which is a powdered gypsum that is mixed with water into a paste. Apply the joint compound along each joint, forcing it into the crevices. Don't try to fill the joint on the first application. As soon as the joint compound has been applied, run a strip of perforated reinforcing tape the full length of the joint, imbedding it in the compound. Let the compound dry before going further.

After the compound is dry, apply a second coat of joint compound, using the wide-blade knife. This will level the compound almost flush with the surfaces of the butted panels. At the same time, fill the dimples left by the hammerhead.

When the final coating of compound has dried thoroughly, go over the joints with medium-fine sandpaper and a sanding block to smooth them. The trick in sanding is to bring the compound to an exact level with the surface of the panel without marring the paper at the sides of the joints. Allow the compound to dry thoroughly before painting.

INSTALLING GYPSUM BOARD

1. Reduce the board to the desired length by scoring on the paintable side, using a straightedge and craft knife; then snap off the excess.

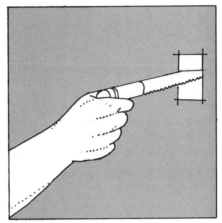

2. Mark and saw outlet box cutouts.

3. Install ceilings first; use a T-brace to position the panels and hold them while you nail.

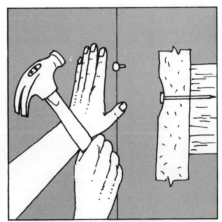

4. Nail along each long edge with coated or blued wallboard nails about 8 to 10 inch OC. Put two or three nails in each panel's center area; with the last stroke, form a "dimple" with the hammerhead to set the nail below the surface level.

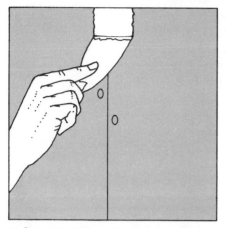

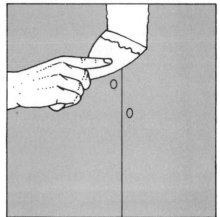

5. Cover the "dimples" with a dab of joint compound and level with a joint knife.

6. Apply joint compound along all nailed edges, but don't fill the trough created where their edges meet yet.

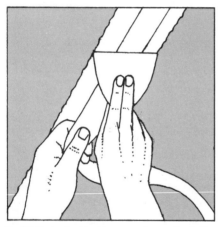

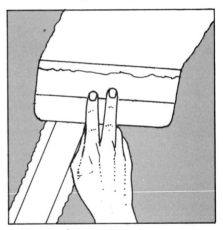

7. Apply perforated tape as shown the full length of each joint, sinking it into the compound.

8. Fill the trough with joint compound and smooth with the broad joint knife; try to finish the surface level with the panels to minimize sanding.

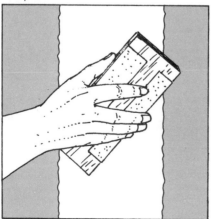

9. Allow the joints to dry overnight, then sand with fine sandpaper and a block.

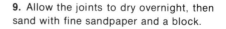

FINISHING CORNERS

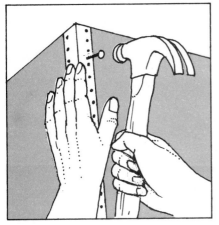

1. Nail metal bead to the outside corner.

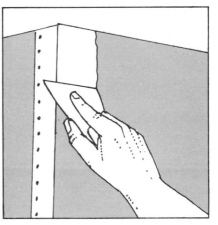

2. Apply a coat of joint compound to cover, leveling the compound just below the surface of the wall.

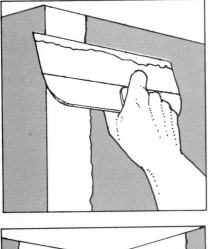

3. Apply a second coat of joint compound with a wide knife, leveling it to the edge of the bead and the wall, then sand as needed.

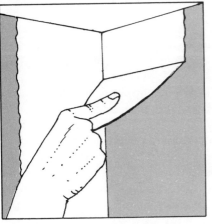

4. To finish an inside corner, apply a thin coat of joint compound in the corner and about 2 inches over the wall surface.

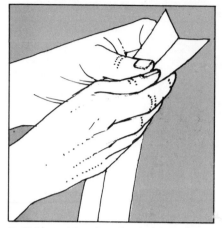

5. Fold a strip of perforated tape long enough to cover the entire joint.

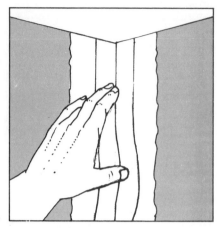

6. Press the tape into the joint compound with fingertips.

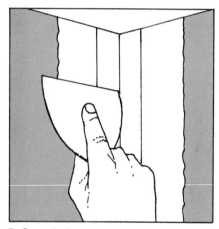

7. Smooth the tape into compound with a joint knife.

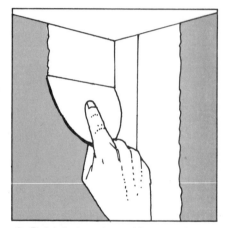

8. Finish by tapering a thin coat of compound into the corner with the wide joint knife, let dry, and sand.

Finishing gypsum board. Any type of paint can be used, flat, semigloss, glossy, or textured. The job can be done with either brush or roller. Two coats will usually be required. Light-hued paints generally hide the joint lines where compound has been applied. If a dark tone is to be used, the joints should be sealed with thinned sealer or with a thinned coat of undercoat.

Most professional installers use joint compound and tape to finish both inside and outside corners. However, outside corners in traffic areas where they are subjected to accidental bumps by furniture being moved or youngsters playing indoors on rainy days can be finished with a metal beading. Nail or staple the beading in place before taping, and apply joint compound over the beading. Level the second coat of compound with the surface of the wall by drawing the joint knife down while resting one side of the edge on the beading, the other side on the surface of the panelboard.

When installing drywall in a moisture-prone room such as a bathroom, kitchen, or laundry room, water-resistant panels are indicated. The gypsum core of this type of paneling contains water-resisting materials, and the covering is also treated. This type of gypsum panelboard is also used when tile is to be applied on the wall.

PREFINISHED PANELS. In addition to prefinished plywood paneling, which was discussed in Chapter 3 as an interior wall covering, there are prefinished gypsum-board and hardboard panels that can be used for this purpose. There are also several methods of applying each of these products, depending on whether you are renovating an existing room with solid walls or one with badly cracked walls, or using them in a room addition or new construction. Let's look at each of these products individually before getting down to installation details.

Back in Chapter 3, it was noted that there is a category of plywood which actually should be placed in the panelboard group. This is a very thin, flexible plywood, $^1/_8$ to $^3/_{16}$ inch thick, with an outer face of prefinished wood veneer. The chief difference between this plywood and the decorative panels of hardboard and gypsum board is that the plywood is 100% wood, while the woodgrain patterns on hardboard and gypsum-board panels are either lithographed on the surface or applied in the form of a vinyl film overlay that is bonded to the panel.

There is also an extremely thin finished hardboard paneling, approximately $^3/_{32}$ inch thick, with woodgrain or textured patterns lithographed on one surface. Both the ultrathin plywood and the ultrathin veneer panels are designed to be applied with an adhesive or mastic over sound existing walls or over new walls of gypsum board, hardboard, or particleboard. These two products are not suitable for installation over furring strips spaced 16 inches on center, but can be installed over furring that is more closely spaced.

Prefinished hardboard paneling is available either in woodgrain finish or in a marble-pattern finish; the latter is called tileboard and is intended to be used primarily in bathrooms. There are also patterns and solid colors which are suitable for finishing a kitchen or a laundry room. These panels can be installed in several ways. They can be applied to sound existing walls with adhesive, put over badly damaged walls by using furring strips, or put on new walls that have a subwall of hardboard or gypsum board. Metal corner and border strips as well as edge strips in compatible colors are available for this latter method of installation. So are color-matched nails for attaching the panels to walls or furring strips; the nails can be used alone or in combination with adhesives.

Prefinished gypsum-board panels are made with four standard wood-grain patterns and colors bonded to one face in a vinyl film. The vinyl material used on these panels is also available with a cotton backing to be applied to doors if you want a perfectly matched wall which is broken by a door. Prefinished gypsum-board panels have angled edges which give a V-groove at each joint when installed with adhesive, or the panels can be fitted in metal channels that are finished to match the panels. These channels are available in all the forms necessary to hold the panels in place and to hide their edges. There are dividers into which adjoining panels are butted, inside and outside corners, edging, and cap strips.

Installing prefinished panels. Each kind of prefinished panel described in the preceding paragraphs, with the exception of the ultrathin types, can be applied to walls by several different methods. The ultrathin panels tend to bulge unless they are put on with adhesive, and they must be installed over walls that are sound except for minor cracks. Even when applied to furring strips spaced much closer than the customary 16 inches on center, these thin panels may develop bulges, though the plywood panels are less inclined to bulge than are those with an extremely thin hardboard base.

All the other types of panels can be applied using your choice of methods: adhesive over existing walls, adhesive over furring strips, adhesive and color-matched nails, nails alone over existing walls or over furring strips, or metal channel fittings.

Tools. Regardless of the application method you choose, your tool list will be pretty much the same. You will need a fine-tooth crosscut saw or a power saw (either a circular saw or a saber saw) with a fine-toothed blade, and a keyhole saw for cutting utility-box openings and for cutting around windows and doors. In addition, you'll require a hammer and nail set, a spirit level or plumb bob, a rule or tape, and a caulking gun and a padded block if you intend to use adhesive.

A push drill or power drill will be needed for predrilling nail holes in hardboard, and it may on occasion be necessary to drill for a nail in metal molding when the end of a panel falls between the factory-drilled holes. A chalkline will save a lot of pencil marking.

When working with all types of prefinished paneling, take great care not to mar its surface with the tools you use. In sawing, turn the face of the panel up if using a handsaw, down if using any type of power saw. When using a handsaw, cut only on the downstroke. When using a portable power saw, a wise precaution is to apply a strip of masking tape over the saw line and then transfer the saw line with a straightedge to the tape if the tape isn't translucent enough to allow you to see the line under it. This minimizes splintering when cutting plywood and edge fuzz when cutting hardboard, and reduces the tendency of the cardboard covering of gypsum board to roll up along the outline.

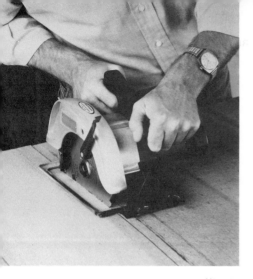

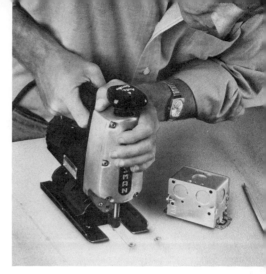

Masonite

Cut panelboard face down if power saw is used, but cut it face up if you do the job with a handsaw.

Use a saber saw or keyhole saw to cut holes for electrical outlet boxes; observe the same procedure mentioned earlier.

If you have wall preparation to do when applying the panels directly to an existing wall, large cracks or damaged areas should be filled with patching plaster and troweled smooth, or sanded. The tools listed above will serve all your needs in applying furring strips as well as the panelboards.

Furring strips. These may be 1×2 or 1×3 battens. Using furring strips presupposes that the wall over which the paneling is to be placed is in bad shape, so be sure the furring strips are set plumb in relation both to each other and to the floor and ceiling. If you fail to plumb them to the floor and ceiling your finished wall may look like an inside-out version of the Leaning Tower of Pisa, and if you fail to plumb them in relationship to each other you'll wind up moving them or scabbing nailing blocks onto their edges, as the panel joints won't butt over their surfaces.

Remember that not all walls are necessarily plumb and that not all room corners form true right angles. Check the room before you begin installing the panels; you might find that one of the corners has gotten out of true because of foundation settling, or because it was framed by a wood butcher disguised as a carpenter. When you encounter deviations from true straight lines, or when you find a wall that leans in or out at top or bottom or has developed inward or outward bulges, use shims behind the furring strips to adjust their levels as needed.

Be very sure that a majority of the furring strips are nailed to studs and that the strips fall at the proper 4-feet-on-center spacing so that the panel edges can be butted over the strips.

Applying panels with adhesive. As a general rule, any adhesive or mastic suitable for applying one brand or type of prefinished paneling is suitable for all brands or types. If in doubt about this, check the specifications and recommendations of the manufacturer of the panels.

If the wall surface over which prefinished paneling is to be installed is not in good shape, or if extra insulation is wanted, place furring strips 16 inches OC along the walls, over existing studs, if possible, and at top and bottom. True the strips with a level, not by eye alone.

Masonite

Most adhesives or mastics are formulated to retain a limited degree of flexibility. In other words, they never really become rock-hard; they cure rather than dry. In spite of the seeming solidity of a room's walls, they are always moving imperceptibly. When the atmosphere is exceedingly humid the walls will expand a bit; when conditions are dry or when a room is heated, a small amount of contraction takes place. Foundations shift to a minute degree, and the walls follow these shifts. Because of this, panels must be applied to a wall in a manner that lets them retain a limited flexibility so that they will give that microscopic fraction of a millimeter when it's necessary.

This is why adhesive should be applied with a caulking gun rather than a brush, and it's also why the panels should never be forced into place. For safety, leave about $1/8$ to $1/4$ inch free space between the top edge of the paneling and the ceiling, and a gap of up to 1 inch or $1^1/2$ inches at the bottom. The top gap will be hidden later by a molding, the bottom gap by a baseboard.

Run a wavy bead around the perimeter you've marked on the wall or furring strips to indicate the panel's edges. You will, of course, have trued up the panel and positioned it precisely, and marked its edge line on the wall with pencil or a chalkline. The adhesive line should be about $3/4$ to 1 inch inside the marked edge line. Follow this with a series of beads inside the area that will be covered by the panel. These beads can be in any pattern that pleases your artistic fancy: straight vertical or horizontal lines, an X or a Z pattern, chevrons, slant marks. The beads should be about $^3/_{16}$ to $^1/_4$ inch in diameter. (See photo sequence page 118.)

Ultrathin paneling and some of the thin hardboards should be applied using the cohesive method: the mastic or adhesive is applied to both the wall or furring strips and to the back of the panel. The patterns on the wall and those on the panel should be approximately the same, so that the two beads of adhesive will meet and merge as they spread when tapped.

Now, position the panel. Have a couple of nails ready and a hammer where you can reach it. When the panel is in perfect alignment with its positioning marks, tack it in place with a nail on each side. Don't drive these nails all the way; they will be pulled out later, and the holes left can be filled with wood putty or a wax stick that matches the panel's finish.

Press the panel firmly against the wall, then go over it with a padded block and a hammer, tapping it firmly to spread the adhesive and flatten the beads. Most of the adhesives used in panel application set or cure within two hours. After the first panel has been set, leave the nails holding it in place while you set the adjoining panel. Allow a very small space—the thickness of two sheets of paper—between the panels, to allow space for normal expansion due to humidity. Don't force the edges together or buckling may result.

When you've gone all the way around the room, remove any tack nails that may have been used to hold the panels in place temporarily, and fill the nailholes with matching wax-stick or putty materials; the dealer from whom you bought the panels will have these matching sticks in stock. Just rub the tip of the stick over the nailhole until it is filled, then wipe away any excess mending material that may have rubbed off on the surface around the filled hole.

If any of the mastic oozes out along the joints, wipe it away with a clean dry cloth. If the adhesive smears, consult the label for the proper solvent with which to remove it. Usually, this will be mineral spirits (paint thinners), but some adhesives call for other cleaners. Use the least possible amount of the fluid on a clean cloth, and after the smear has been removed, wipe the area with a dry cloth. Some cleaners leave a residue that if not cleared away may attack the panel's finish.

This method is used when you use adhesive alone or adhesive and nails. Color-matched nails that blend almost unnoticeably with the surface of the panel are available from the dealer who handles the paneling. Predrill for nails in hardboard panels, and drive them with a nail set to avoid chipping off the finish. If a speck or two flecks off the nailhead, use the wax or putty stick to recolor it.

Applying panels with nails. Although the best method of applying prefinished paneling is with adhesive or adhesive and a few widely spaced nails, you can use nails alone. If you choose color-matched nails,

APPLYING PANELS WITH ADHESIVE

1. Position the panel carefully and mark placement lines on the furring strip.

3. Push the panel against the furring strip and move it up and down and from side to side to spread the mastic, then pull the panel away from the strip and hold it for about 30–45 seconds to expose the mastic to air.

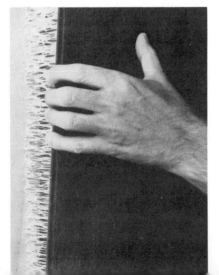

2. Apply beads of mastic along the placement line on the furring strip and other beads that will fill in the voids along the edges of the panel.

4. Position and level the panel; you have approximately two minutes working time before the mastic sets up.

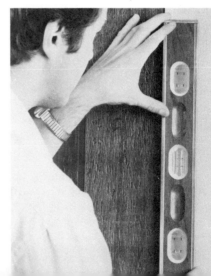

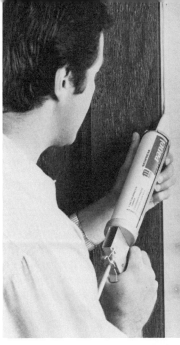

6. Lay down a sparse solid bead on the next two furring strips.

5. As a precaution, drive one or two finishing nails into each edge of the panel; this will prevent any slipping when you are butting the next panel against it. The first panel is the most critical one because if it isn't level, every other panel will be off-line.

7. Butt the next panel into place, move it up and down and side-to-side to spread the mastic, pull away, position.

8. To save time, instead of holding the panel while the mastic is setting up, drive a nail or two through the edges to insure full adhesion and proceed to the next.

All photos courtesy Masonite Corp.

119

they will not be noticed. If you elect to use plain nails, use finishing nails, predrilling for them if hardboard paneling is being applied, and after countersinking the heads, fill the holes with a wax or putty stick applied as described in the foregoing section.

Space nails about 8 inches apart along the joints and drive two or three into the studding in the center area of each panel.

Applying panels with molding. There are two types of extruded metal molding used in installing prefinished wall panels. One type supports the molding, the other simply covers the bare edges at joints and corners. The former must be attached to the studs or wall or furring strips with small flathead nails; these are usually furnished with the molding, and the molding is predrilled to accept the nails. The second type simply slips over the edges of the panels, which are applied with adhesive, to trim the joints and finish corners so that no bare edges are exposed.

Usually, when using either type, you will want to apply molding around the entire perimeter of the panel; these moldings allow you to apply the paneling flush with the ceiling and floor without having to install quarter-round or cove along the ceiling line or baseboards at the bottom. An installation of this kind requires that you miter the corners of the molding. The best setup I've found for cutting miters in this kind of molding is the miniature miter box and razor saws made by X-Acto. The combination allows for very precise cuts and is easy to handle. Trying to cut small metal molding in a standard miter box with a hacksaw can be a real headache.

If you're using nail-on molding, you'll have to snap hardboard panels into place, and the bottom molding strip must be put on with adhesive. The slip-on moldings that attach only to the panelboard edges present no real installation problem. The divider strips that go between the joints can be applied to the molding already in place, then the top and bottom strips are put on the piece being butted in, and it is slid into the molding. The strip on the opposite edge is applied after the panel is in position. Miters must be prefitted, of course.

If you want to have neat corners without the time-consuming job of trimming and fitting, there are edge and corner moldings that can be used. These don't need to be fitted to the panel edges, but are applied conventionally with nails after all the paneling has been installed.

Special applications. Tileboard, a hardboard paneling with a marble-pattern or high-gloss-enameled surface finish, requires another procedure. This material is designed for use in bathrooms, kitchens, and laundry or utility rooms, where excessive internal moisture is often created. Its installation requires special attention.

Ideally, tileboard should be installed over a level wall that is in sound condition. The wall should be a minimum of $3/8$ inch thick. This is especially important when making the installation in tub and shower

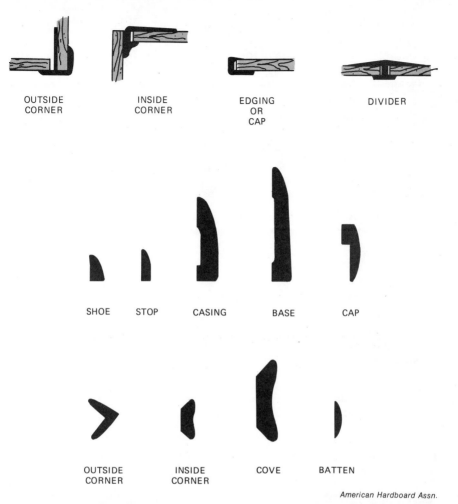

OUTSIDE CORNER INSIDE CORNER EDGING OR CAP DIVIDER

SHOE STOP CASING BASE CAP

OUTSIDE CORNER INSIDE CORNER COVE BATTEN

American Hardboard Assn.

If you prefer to install hardboard with metal strips, the top row shows the shapes available in colors that match hardboard panels. Trim strips of wood for framing door and window casings and for use as baseboards and ceiling coves are shown in the two bottom rows.

enclosures, in washer and drier niches, or on a wall where a dishwasher is to be used. If the old wall in such areas is in very bad shape, you might consider applying an inside sheathing of hardboard, particleboard, or triple-sealed gypsum board. These can be placed on the old wall or on studs or furring strips. A vapor barrier of polyethylene should be installed between the old wall and the sheathing, which in itself will provide a partial vapor barrier.

Tileboard must be installed in much the same manner used in applying vinyl flooring and countertops. Both wall and panel must be coated with adhesive and the adhesive raked with a serrated spreader. Molding should be installed at each joint and corner, and around tub- and shower-enclosure edges. In cutouts made for pipes, a good grout—latex or silicone—should be used to fill the voids between pipe and the rim of the cutouts. The entire area of the room, including the ceiling, should be paneled.

All the techniques for carrying out these jobs are pictured here, or have been pictured earlier, as have the methods for measuring and installing the tileboard.

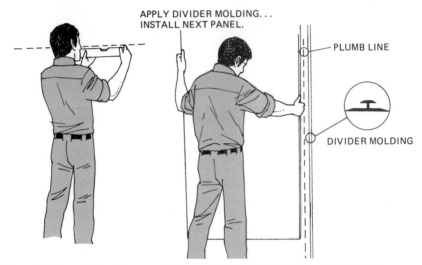

When installing tileboard, use a spirit level to establish a true vertical angle for divider moldings.

Tub or shower-stall joints and cutouts made to accommodate plumbing fixture, pipes or faucets, should be completely filled with latex or silicone caulk after the tileboard is in place and before the fittings are replaced.

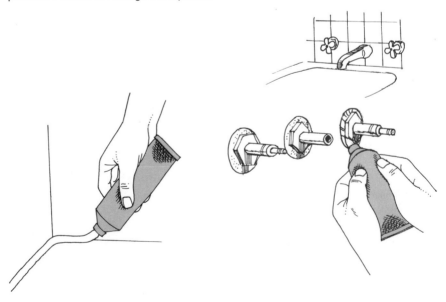

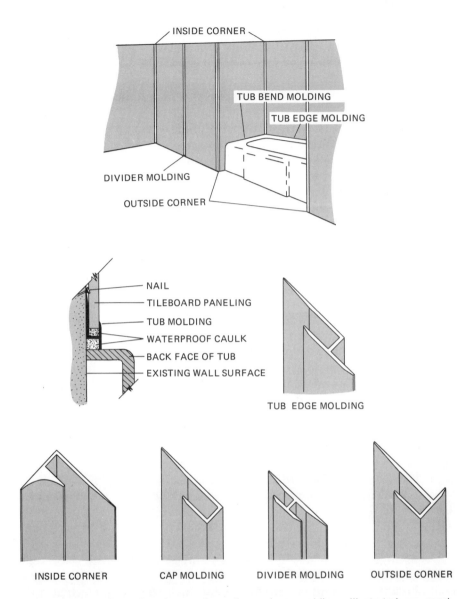

These sketches show the places where the various moldings illustrated are used. Tileboard should be installed with molding to give the best service life.

Vapor and sound control. In order to minimize the tendency of any hardboard paneling to warp when attacked by a condition of alternating humidity and drying, a wise precaution is to install a vapor barrier. This is practical only if the installation is being made over furring strips—and keep in mind that any deterioration an existing wall may have suffered is very likely the result of changes in humidity. If you plan to install paneling over an existing wall, you should certainly ask yourself whether the existing wall has a vapor barrier, and, if it does, whether the barrier has lost effectiveness with the years. However, you cannot install an effective vapor barrier over an existing wall if you plan to put the hardboard panels up with adhesive.

Polyethylene film and vinyl film are the most commonly used vapor barriers. However, neither polyethylene nor vinyl film will bond to adhesives. If you want to install decorative hardboard panels with adhesive and don't wish to put in furring strips, you can use triple-sheathed hardboard panels, which form a very effective vapor barrier, nail them to an existing wall, and then apply the finish paneling with adhesive direct to the sheathing.

Sound control is a real asset to any home. It is easiest to achieve with new construction, but you can create an effective sound barrier between rooms if you're willing to sacrifice a few inches of their size.

Look on any drywall-finished partition as a kind of drum. It resonates in response to sound, just as a tightly stretched drumhead resonates when tapped with a stick. A drum will also transmit music or voices by resonance. Your objective is to dampen this tendency any flat stressed surface has to respond to airborne sound waves.

In existing construction, you will have to provide an additional bit of space into which sound-dampening insulation can be placed. Do this by going out from the existing wall with a new set of studs, allowing enough space between these studs to install batts of insulating material. The thicker the batts, the more effective they will be in arresting sound vibrations. The batts should be compressed as much as possible between old and new walls to minimize the amount of space lost by adding the false wall. Then, apply hardboard paneling over the new studding.

In new construction, very effective sound dampening is much easier to achieve and requires the loss of only $3/4$ inch of space from one of two adjacent rooms. Widen the 2×4 plates and headers. On load-bearing partitions, space 2×4 studding on conventional 16-inch centers flush with the wall edge of the plates and headers, and on the opposite wall edge set the first 2×4 stud 24 inches from the corner flush with the edge, then continue with a second set of studs on 16-inch centers along that wall edge.

This method of alternate-flush setting makes space between the studding for the installation of continuous insulation batts, and removes any connection between the studs that support the paneling in opposite wall surfaces. There is no connection between the two skins of the drumhead, and the insulating batts prevent sympathetic resonances from being transmitted.

Finishing. When using prefinished panels, your only finishing job is filling nailhead holes with colored stick wax or stick putty, or touching

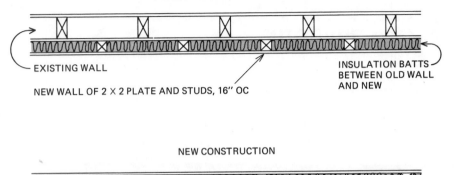

EXISTING CONSTRUCTION

EXISTING WALL

NEW WALL OF 2 × 2 PLATE AND STUDS, 16" OC

INSULATION BATTS
BETWEEN OLD WALL
AND NEW

NEW CONSTRUCTION

Methods of achieving sound control in rooms finished with panelboards are illustrated above.

up the heads of color-matched nails to restore spots that have been flaked off by the hammer.

Stick fillers of one type or the other are sold by the supplier from whom you bought the panelboard, in shades that match the color of their finished surface.

6 | Working with Laminates

FIRST, LET'S MAKE sure we're all thinking about the same product. There are a number of laminated panels used in building and home crafts or interior decoration. Plywood is a laminated product, and so is gypsum board. So are main support beams in many wide-span structures. But what this chapter is concerned with are thin semirigid panels which most of us think of as being all man-made plastic, but which are actually made from layers—*laminae*—of a wood product, kraft paper. This ordinary brown wrapping paper, produced from wood pulp, is the base or core of all laminate panels, even though the paper is unseen and its presence unsuspected by many who use these decorative panels.

Laminate panels grew from an unlikely source. The use of phenolic-impregnated kraft paper, which is the base of panel laminates, started in the early days of the automobile. As cars grew more complex, their electrical systems followed suit. An auto's electrical wiring soon became much more than a pair of leads from battery to generator and a set of wires to headlights and taillights. To operate gauges and accessories, the current had to be distributed from a central source, a panelboard which could be equipped with clips and fuses and resistors to deliver a multitude of different wires carrying several varieties of currents to their destination.

Metals, of course, were out of the question, because they were carriers of current. So was wood, which deteriorated and split under the exposure these panels received. Sheet plastics of the period cracked and broke under vibration, and rubber insulating washers that mounted metal fittings on a metal plate quickly deteriorated under exposure to lubricants and weather.

Laminating several plies of cloth in a phenolic-resin base and fusing the plies under heat and pressure produced a tough, impervious panel with very high insulating value. The first of the laminate panels were made with cloth layers; those old enough to recall their initial appearance will remember that no matter what the color of the laminate, it had a perceptible woven pattern imparted to it by the cloth.

At first, laminate panels were much thicker than those in use today, and their use was largely confined to commercial establishments such as restaurants and soda fountains, where they replaced older countertops and tabletops of heavy marble or enameled metal. The first patterned laminates, on a kraft-paper base, were produced in an effort to overcome the resistance of café and soda-fountain owners to any topping material that did not have the prestigious appearance and smoothness of marble.

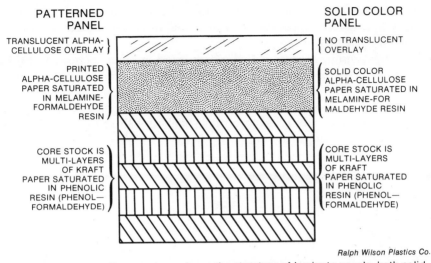

PATTERNED PANEL

TRANSLUCENT ALPHA-CELLULOSE OVERLAY } — { NO TRANSLUCENT OVERLAY

PRINTED ALPHA-CELLULOSE PAPER SATURATED IN MELAMINE-FORMALDEHYDE RESIN — SOLID COLOR ALPHA-CELLULOSE PAPER SATURATED IN MELAMINE-FOR MALDEHYDE RESIN

CORE STOCK IS MULTI-LAYERS OF KRAFT PAPER SATURATED IN PHENOLIC RESIN (PHENOL—FORMALDEHYDE) — CORE STOCK IS MULTI-LAYERS OF KRAFT PAPER SATURATED IN PHENOLIC RESIN (PHENOL—FORMALDEHYDE)

SOLID COLOR PANEL

Ralph Wilson Plastics Co.

Greatly enlarged, the diagram above shows the structure of laminate panels, both solid-color and patterned.

Today, of course, laminate panels in many textured surfaces are widely produced and used.

HOW LAMINATE PANELS ARE MADE. Several different kinds of plastics go into the surfaces of modern laminates. The most common are melamine and urea formaldehyde, produced from calcium cynamides. The line between all the plastic resins used in the surfaces of laminate panels is thin and wavering to anybody except an industrial chemist, and for all practical purposes as far as the end use of laminated panels is concerned, the differences are fewer than the similarities. All the materials used in making panel laminates result in a product which has the same basic application, is worked by the same methods and with the same tools, and has approximately the same qualities, both internal and surface.

Universally, in the process of manufacturing laminate panels, kraft papers are impregnated with a phenol-based liquid as the paper goes through a series of troughs between rollers. As the last steps of the process approach, the layers (remember, *laminae*?) of papers are squeezed together and coated with a film that bears the surface decoration. The layers then go through presses to compact them into a homogeneous mass and travel through ovens. As the chemicals used in the impregnating of the layers are all thermosetting, the wide ribbon of laminated material dries and hardens into a glass-smooth ribbon. The ribbon is sawed into sheets, or panels, and cured under controlled temperatures.

This is the basic process, and any other treatment, such as surface texturing, becomes an accessory. The underside of the laminate panel is either given a slightly roughened texture to increase its adhesive-holding ability as it goes through the final pressing, or is sanded after curing to roughen it for this purpose.

What is produced, as those who have used laminates know, is a thin panel, semirigid, of a very dense, very tough material which is highly resistant to normal wear, such as scuffing or abrasion, is quite moistureproof (though still subject to tiny dimensional changes due to humidity), and, when bonded to a thick backing called the *substrate*, is very resistant to impacts.

In the first step of laminate manufacture, kraft paper passes through troughs which saturate it with a phenolic resin, then through rollers which remove excess liquid.

Ralph Wilson Plastics Co.

A special type of alphacellulose paper bearing a lithographed or printed pattern passes through a melamine bath and is overlaid on several layers of the resin-saturated Kraft paper shown in the preceding picture.

Ralph Wilson Plastics Co.

In giant presses, the layers are fused under heat and pressure to form a glass-smooth, highly durable thin panel.

Formica Corporation

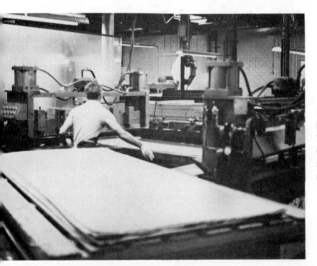

From the presses, the laminate panels go through an edger which trims them to precise dimensions and sands the back side to give them better adhesion capability.

Ralph Wilson Plastics Co.

SIZES AND PATTERNS OF LAMINATE PANELS. Eight major manufacturers produce the bulk of the laminate panels sold in the United States. While most of them produce panels in standard sizes, the number of sizes varies from one maker to another, as do the number of patterns available and the number of panels made for special-purpose applications. Regardless of the brand name on the panel, if the manufacturer is a member of the industry's national trade association you can be sure of its quality, for the panel will have been manufactured to conform to industry standards.

Checking the products lists of some of the leading makers of brands such as Formica, Micarta, Wilsonart, Consoweld, Nevamar, Fabricon, and others, a typical cross section will show that a manufacturer's product line will be available in three thicknesses and four to six widths, and in lengths from 6 to 12 feet. However, the 4×6-foot or 4×8-foot panel module is still the size stocked by all but the largest retailers. On special order you can get panels 24, 30, 36, and 60 inches wide, and lengths of 10 and 12 feet. It is a normal practice to cut laminate panels slightly in excess of nominal dimensions in order to allow for edges to be finished to actual dimensions after the material is installed.

Each thickness of laminated paneling bears a relationship to end use rather than stress factors, as is the case with other paneling; these panels are not classed as structural materials in the sense that lumber, plywood, and hardboard and other panelboards are. Actually, laminate panels are governed by two thickness standards, one being the thickness of the face or decorative surface, the other of the backing.

General-purpose laminates are nominally .046 inch in total thickness, and their surface thickness is slightly greater than 1% of this total. Panels intended for wall application are slightly thinner and have a shallower face. Panels for fire-resistant applications are slightly thicker, and their face surface is deeper. These differences are expressed in hundredths of an inch, of course, and for general home use you can safely use any of the general-purpose or special-purpose panels anywhere that pleases you, with one exception.

This is a special-use laminate designed for use as door paneling. It has half the overall thickness of the general-utility laminates, but with a surface thickness equal to the general purpose type. The reason, obviously, is to save weight. The minute details of laminate thickness is a matter of importance primarily to the commercial builder who must quote specifications or meet specifications in contract bids, and to the fabricator of laminate-covered countertopping-splashboard-basin modules, whose production-line operations use machines to form a continuous covering that takes in splashboard, top, and edge.

This type of application is called postforming, and requires the use of both a specially manufactured laminate and a special machine. The laminate has creped kraft-paper coring which allows it to bend readily as the postforming machines heat the material and press it into the contours of the prefabricated backing. It is not a style of application that can be done in the home workshop. However, the postformed countertop units are sold by many retail dealers for installation by home craftsmen. We'll look at the installation details of these in a later section of this chapter.

Surface finishes—smooth and glossy, smooth and satiny, patterned and textured—vary so widely among manufacturers that a small book could be filled just listing them and their descriptions. In the broadest general terms, there are solid colors, smooth patterns, textured patterns, woodgrain patterns, metallic finishes, and textured finishes.

A typical catalog of one large-scale manufacturer lists more than thirty solid colors, more than twenty woodgrain finishes, a dozen decorative-design finishes with smooth surfaces, another dozen with textured surfaces, a half dozen textured surfaces designed to resemble stone or some special type of wood, and still another small group in metallic finishes. Some manufacturers also include a special type of laminate in several finishes designed for use on countertops and tabletops that has a layer of thin metal foil under its surface finish, to carry away the heat of a carelessly dropped cigarette or cigar and prevent the finish from being marred by the heat.

If you're prepared to do a bit of catalog-scanning and shopping, you can find quite literally any finish you might reasonably desire in the laminate field.

WORKING WITH LAMINATES. Laminate panels can be worked with either hand or power tools. The hand tools needed are a crosscut saw with 12 or 10 teeth per inch, or a mini-hacksaw with a standard blade, or a glasscutter or special laminate scoring knife; measuring equipment, including a rule or tape and a square, and a compass if curves must be cut; a paint roller; a roller or mallet and padded woodblock; and a smooth-cut file or coarse and fine sanding blocks.

Power tools should include a circular saw with a tungsten-carbide-toothed blade or a saber saw with a metal-cutting blade; and a router. When using power tools you will, of course, also need the measuring and applying tools listed above.

Surface preparation. First, let's look at the types of surfaces over which laminates can be applied. They include solid surfaces made from any kind of lumber, plywood, particleboard, or hardboard. Plaster and gypsum-board surfaces are not suitable for laminate application. The glues used in fastening the laminate to these types of surface may soften the surface and create problems of installation as well as of permanent adhesion. You can also apply laminates to metal surfaces by using contact cement. You can apply new laminate surfaces over old ones if the old surface is sound and free from bubbles or loosened areas where it has separated from its substrate.

Any surface on which laminate is to be applied must be smooth and even, free from large dents, loose material of any kind, oil or grease, and, in the case of metal surfaces, rust. The best surface preparation is a good sanding with a finish sander. A belt sander can be used to remove bumps or other surface flaws, or to smooth areas that have been plugged or patched, but an oscillating or straight-line power sander, or hand-sanding with a block, should be the final step. After sanding, all dust should be removed by wiping the surface with a damp cloth or a cloth moistened with paint thinner.

Cutting and trimming. There is one precaution which must be taken whether you are cutting laminate with a handsaw or a power saw: the panel must be supported on both sides of the sawline, and as close to the line as practical. Remember, you're working with a very thin material, and one which may crack under the vibration created by sawing. Use a board under each side of the saw, making sure both boards extend the entire length of the cut.

When you use a handsaw, the panel should be placed face up; when either a portable circular saw or a saber saw is used, face down. A strip of masking tape should be centered on the sawline, and if the tape hides the line, mark a new line on the tape. Apply the tape to both sides of the panel. With a handsaw, cut only on the downstroke. Advance a power saw slowly and steadily.

Cutting straight seam lines when sections of laminate must be butted together is done most satisfactorily with a router. A small-diameter carbide veining bit is the best cutter to use. Support the work from below, just as when cutting with a saw, and clamp boards on top of the work, spaced to accommodate the base of the router. Overlap the pieces to be joined, making sure the long edges are in alignment. Pass the router along the cutline, cutting a matching edge on both pieces at a single pass.

Always measure carefully the surface to be covered and then allow an overhang on all edges; this will vary according to the method you will use in the final finishing steps after the laminate has been bonded to its base. If you do not plan to use a router to finish the edges, allow an

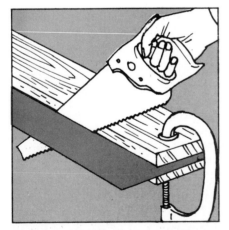

No matter what type of saw you use to cut laminate, be sure the panel is supported at top and bottom as shown here. When you saw with a handsaw—10- or 12-tooth crosscut—the face of the panel should be up. With a portable power saw, the face should be down.

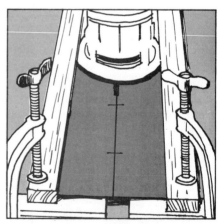

Cutting edges that must butt together is most easily done with a router. Clamp the two sheets as shown, overlapping them about 1 to 2 inches, and use a small-diameter carbide veining bit. You can use this technique with a saber saw, but remember to place the sheet face-down and clamp the overlapped edges.

overhang of $^1/_8$ to $^1/_{16}$ inch, and do the final finishing with a file, followed by sandpaper. Protect the side with a strip of masking tape when filing or sanding to avoid scratching it.

If you will be using a router, the amount of overhang can be as much as $^1/_4$ to $^3/_8$ inch, and you should use a carbide edging bit, which is self-guiding. These bits have a bearing below the cutting edge which guides the cut. Smear petroleum jelly along the side to avoid scoring or scorching the surface with the roller-bearing guide. Router bits are available to make both straight and miter-edge joints.

When you must drill holes in laminate which has not been bonded to a base, always use a scrap of wood for a back-up block under the drill bit. Clamp the laminate to the wood, protecting the surface where the clamps touch it with masking tape. Standard wood bits can be used in a handbrace, metal bits in a power drill. Always prick the surface with a sharpened nail or punch to make a starting dent for the drill.

In kitchen or bathroom installations, where cutouts must be made for pipes, use an expansion bit in a brace or a hole saw in a power drill. Large holes can be cut with a router, using a template, or with a fine-tooth keyhole saw. However, if you are making the cutout for a basin or sink, this should be done only after the laminate has been bonded to its substrate. Use a keyhole saw or router; you'll be running a risk if you use a saber saw, no matter how fine the blade. The reason for turning the back of a laminate panel up when cutting with any portable power saw is to avoid chipping; portable power saws cut on the upstroke, and their teeth are more likely to tear a chip from the laminate on emerging from the surface of a panel than when entering it.

Laminate panels can be cut with a glasscutter, or with a special scoring blade in a craft knife. Both face and back must be scored, and only straight cuts are possible. You cannot make curved or inside corner cuts by the scoring method. The grooves on both faces should be perfectly matched. There is also a special tool fitted with an adjustable tungsten-carbide scoring point which is designed especially for use on laminates. There is a similar hand tool made for trimming the edges of laminates after they've been bonded to their backing; this device can be adjusted to produce either a square or a beveled edge. To cut laminate panels by scoring them with any kind of the foregoing tools, the panel must be clamped between two boards after it has been scored, with the scoreline parallel to the edges of the board and a gap of between $^1/_4$ and $^1/_2$ inch between scoreline and edge. The waste overhang is then snapped off, just as a glazier would snap a scoreline when cutting a glass pane.

Let me insert a word of caution here, though. The score-and-snap technique is one that requires practice. Professionals who work with laminates daily use it and make it look easy, but the pros are accustomed to making the precisely aligned grooves, one on the face, the other on the back, that are required in scoring. The pros have acquired a feel for the material which comes from constant practice. If you've never used the scoring method, or even worked much with laminates, perhaps you should use another method. A saw will do the job in very little more time, and so will a router, with much less risk of damaging a panel.

Generally speaking, you should avoid the use of edged tools such as planes and chisels when working laminates. This is not because the tools won't do the job; a block plane will trim an overhang quite easily, but at the price of a dulled blade. A chisel, used as a shaving tool, will also trim edges, but its blade must be resharpened before it will work wood well again. A file or a sanding block takes a bit more time, but the time consumed is much less than you'd be forced to spend resharpening the blade of a plane or chisel.

Applying. Until the development during the early and middle 1970s of new adhesives in the urea-formaldehyde and polyvinyl-acetate families, and the improvement of the casein and phenol-resorcinol types of woodworking glues, contact cement was generally considered to be the only suitable adhesive for bonding laminates to their substrates. All these types of adhesives are now being used, except where laminate-to-metal bonds are required. Here, contact cement and slow-setting epoxies are the recommended adhesives.

However, a majority of the users of adhesives other than contact cement are large-scale fabricators of laminates, and the work in which adhesives other than contact cement are used is for the most part carried on in factories or large cabinet shops which are equipped with presses that make bonding much stronger than is possible in the average home workshop.

At the same time, new contact-cement formulas have been developed that allow the home craftsman to take a bit more time in positioning the laminate than was possible with rubber-based contact cement, which was the only type available until recently. For home applications, contact cement is still the most foolproof and easiest to use of all the adhesives when applying laminate to a base of wood, particleboard, or hardboard.

Different methods are required for different kinds of substrate materials. On impervious materials such as hardwoods, finished woods, hardboard, and metal, a single uniform coat of contact cement should be applied first to the back of the laminate, then to the bonding surface. Follow the manufacturer's directions in allowing for set-up time; the time between spreading and bonding varies according to the cement's formulation.

When applying laminate to moderately absorbent surfaces such as new plywood, softwoods, and particleboards, two glue coats are usually better than one. Apply a first coat diluted with about 10% by volume of thinner. Makers of contact cements usually manufacture a thinner suitable for their products. The new water-based contact cements can, of course, be thinned with water; again, the manufacturer's directions and proportions should be followed. After the first coat has sealed the surface of the base, apply a coat of adhesive to the back of the laminate, then to the base, and again follow label instructions regarding set-up or curing times before setting the laminate on the base.

A brush is probably the worst applicator you can use when spreading adhesive. A short-nap paint roller is the quickest applicator. Almost as fast is one of the sponge-rubber dabbers used in applying vinyl-based paints. These dabbers cost very little and are available in widths up to 6 and 8 inches, which makes their use just about as fast as a roller and gives a coating that is just as uniform. Another good spreading tool is one of the wide joint knives used in finishing gypsum-board walls.

Your objective in spreading the adhesive on both the substrate and the laminate is to get a very uniform coating, free from ridges or blobs of adhesive. The smoother the adhesive coat, the easier and the more positive the bond will be.

A good precaution is to apply an extra coat about 2 inches wide around the perimeter of both laminate and substrate. This will assure positive adhesion of the edges.

Use a sheet of kraft wrapping paper between the laminate and the base when positioning the work. A good index of the readiness of contact cement for application of the laminate is that the two coated surfaces be dry enough that the paper will not stick to them. Position the laminate, then pull the paper from between the laminate and its base, pressing the two together.

Another method of holding the laminate off the base is to use thin strips of lath. Lay the strips on the base about 6 to 8 inches apart and then put the laminate down on them. When the laminate and base are aligned to your satisfaction, slip the center battens out and press the laminate down on the base. Remove the battens one by one, working alternately from the center to the edges, and pressing as you go.

Rolling is the most satisfactory means of applying the pressure that brings about the final bonding. Professionals use a J-roller, which allows for great pressure to be applied by hand. Almost equally satisfactory is a double roller, which allows you to use both hands along its centerline. However, you'll get good bond if you simply use a rolling pin from the kitchen, or if you go over the entire surface with a padded block and a mallet or hammer, tapping against the base.

Whatever method is used in bringing about the final pressure bond, always work from the center of the surface toward the edges. You will, of course, have seen that the substrate's surface is free of dust and debris before applying the adhesive.

Should a ripple or a bubble develop during application of the laminate, it can usually be eliminated by heating the bulge with a hair blower or an electric iron. Do not use an open flame, such as that from a blowtorch or propane torch. When using a hair dryer, wave the nozzle back and forth over the rippled area or the bulge at a distance of 8 to 10 inches

COVERING A COUNTERTOP WITH LAMINATE

Covering the surface and edges of a countertop or its equivalent with laminate consists of two basic operations: (1) apply a strip of laminate, cut slightly oversize, to the edges and with a router or file bring the strip flush with the surface; (2) apply a panel of laminate, cut slightly oversize, to the surface and with a router or file bring the panel flush with the edges. The following drawings show the steps in the process.

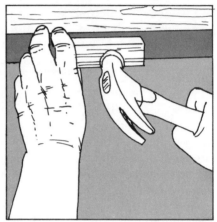

1. Cut a strip of laminate $\frac{1}{2}$ inch wider than the edge. Apply adhesive to the strip and the edge and attach the strip. (See steps 4 and 5 for details on applying adhesive.)

2. Use a small block of wood to protect the laminate, and tap with a hammer along the edge to get good adhesion.

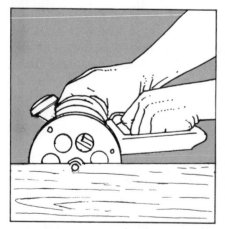

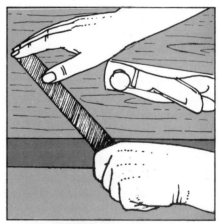

3. Remove surplus laminate with a router, using a carbide laminate edging bit. Smear petroleum jelly along the edge to avoid binding.

If no router is available, cut the strip only $\frac{1}{16}$-inch oversize and finish with a file.

4. After cleaning the surface well, apply contact cement with a short-nap roller to both the surface of the work and the back of the laminate.

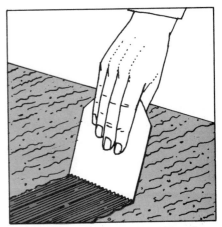

5. Comb the contact cement with a fine-toothed, serrated-edged spreader.

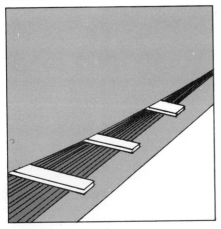

6. Lay strips of board 6 to 8 inches apart on the surface, place laminate in position, then remove strips singly, starting from the center.

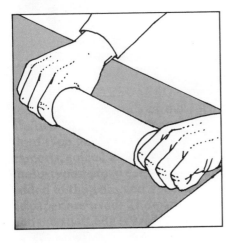

7. Roll the surface from center to edges with a rolling pin (*continued*).

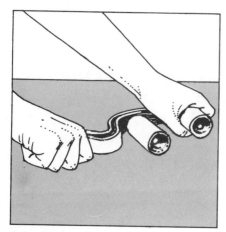

A J-roller does a better job; try to rent or borrow one.

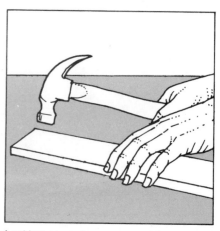

Lacking any roller, use a padded wood-block and tap over the entire surface with a hammer.

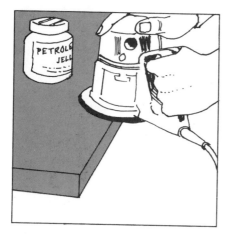

8. Finish the edges of the surface laminate, which should overlap the edge, with a router. Apply petroleum jelly along the routing path.

If no router is available, use a file to bring the surface laminate flush with the edge.

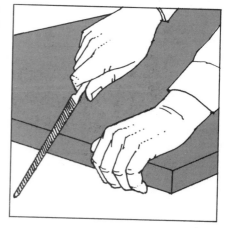

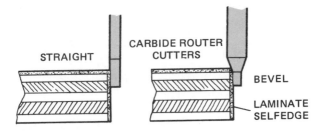

Carbide laminate edging bits come in straight and bevel-cut types.

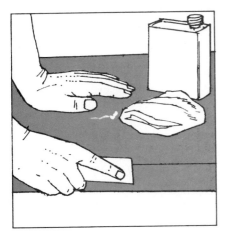

9. Remove any oozed-out adhesive only after the work has been allowed to stand overnight. Use a scrap of laminate as a scraper; it will not scratch as metal tools will. Use lacquer thinner to soften stubborn glue spots, but wipe the surface dry immediately once the spot has been removed.

above the surface and then apply pressure with a roller or padded block, working the bulge toward the edge. You will probably find it necessary to heat the entire path of the ripple as it is forced to the edge.

When using an electric iron, put a single sheet of clean paper between the surface and the plate of the iron. Set the temperature at the lowest range; this is usually indicated by "silk" on the iron. If the paper begins to curl violently and scorch, remove the iron at once. That's an indication you're using too much heat. To be completely truthful, the best way to cure a ripple or a bubble is to avoid creating one in the first place. Eliminating them requires time and careful work, but it is possible to do it.

If you are covering a built-in cabinet, both sides and top, apply the side pieces first. If you are covering a piece of furniture, such as a desk or table, this rule also applies. Work from the bottom up, trim the top edge, then apply the top covering. When you apply the top to a right-angled cabinet, such as might be necessary in covering a kitchen or bathroom counter, there are two equally satisfactory ways of butting the laminate at the corner. One is to cut the two pieces of laminate to form a 45° joint. The other is to carry either of the pieces to the corner and form a 90° joint

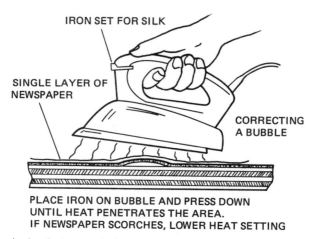

IRON SET FOR SILK

SINGLE LAYER OF
NEWSPAPER

CORRECTING
A BUBBLE

PLACE IRON ON BUBBLE AND PRESS DOWN
UNTIL HEAT PENETRATES THE AREA.
IF NEWSPAPER SCORCHES, LOWER HEAT SETTING

Bubbles or ripples that occur during application of laminates to a substrate can often be removed by using an electric iron at its lowest setting; generally this is "silk."

with the second piece. Both of these are shown in the accompanying illustration. The 45° joint looks neater, but is a bit more difficult to form. The 90° joint will use slightly less material.

Remember that laminates are intended to be applied only on flat surfaces; special hot presses are required to set the material on any curved or spherical base. This does not mean that a strip can't be bent around the curved edge of a countertop or piece of furniture. In this kind of application the base itself is flat, and as long as the curve has a radius larger than 3 or 4 inches, the edge strip will not crack.

Most lumberyards today stock ready-made bathroom and kitchen countertops, complete with backsplash. These are covered with seamless heat-formed laminate, and you can buy one to install yourself, following the steps shown in the accompanying sketches. Or, as you can see from the next set of sketches, you can build your own quite easily. The job's really not all that difficult, and the savings you realize will be substantial. By going the build-it-yourself route you'll also have more options in laminate pattern, backsplash height, and so on.

You can also cover a kitchen door with laminate matching the countertop and the top of a built-in breakfast counter. If you do undertake covering a door, remember that both sides must be covered, for the unequal surface tension created by covering only one side will eventually cause the door to warp. If you don't want to cover the door's other side with decorative laminate, a special laminate backing panel, sanded to accept paint, is available for this purpose.

SURFACE CARE OF LAMINATES. Laminate panel surfaces are not totally immune to damage or marring, though it's a common misconception

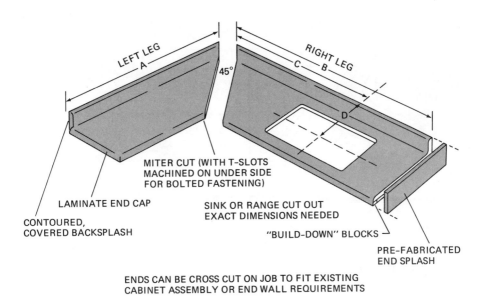

MITER CUT (WITH T-SLOTS MACHINED ON UNDER SIDE FOR BOLTED FASTENING)

LAMINATE END CAP

CONTOURED, COVERED BACKSPLASH

SINK OR RANGE CUT OUT EXACT DIMENSIONS NEEDED

"BUILD-DOWN" BLOCKS

PRE-FABRICATED END SPLASH

ENDS CAN BE CROSS CUT ON JOB TO FIT EXISTING CABINET ASSEMBLY OR END WALL REQUIREMENTS

Most preformed laminate-covered countertops look something like this, except for sink placement. Generally, no cutout has been made; this is part of your job as an installer. Usually, one end is left unfinished to allow you to shorten it.

that they are. While laminate surfaces, when properly bonded to a well-prepared backing, are highly resistant to impact, such as the dropping of a pot or plate, they can be chipped very badly if used as a chopping block or breadboard. They're also quite unkind to the edges of knives.

Laminate surfaces need no waxing or polishing to retain their built-in surface gloss, and can be cleaned with water alone or water with a mild soap or detergent. However, their surfaces are vulnerable to chemical damage. Many of the chemicals commonly found in household cleansers will damage them. Usually, this happens only if the laminate has been cleaned regularly with them, or if spills are not mopped up at once and the surface rinsed with clear water to remove all traces of the spilled liquid.

Here's a list of the chemicals that are harmful to laminate surfaces and the brand names of some of the products in which they occur. The listing of brands isn't complete, but you'll quite probably recognize other products, such as the house brands of major grocery chains, which can logically be expected to contain the same ingredients.

Hypochlorite bleach (Clorox, Hilex, Purex, etc.)
Sodium bisulfite (Sani Flush, etc.)
Lye solutions (Drano, etc.)
Dye (Tintex, Rit, Putman's, etc.)

INSTALLING A PREFORMED LAMINATE-COVERED COUNTERTOP

1. Measure the cutoff and cover the cutline with masking tape to prevent chipping the laminate surface.

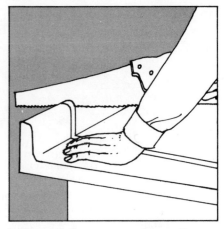

2. Saw off the excess with a 10- or 12-tooth crosscut saw, sawing through the protective tape.

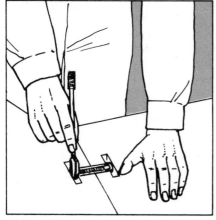

3. Assemble the corner by gluing the joint and pulling its edges together tightly with the special bolt provided.

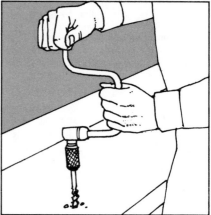

4. Drill corner holes for the sink cutout.

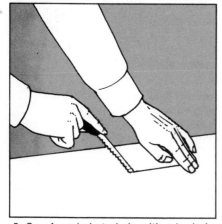

5. Saw from hole to hole with a keyhole saw to remove cutout.

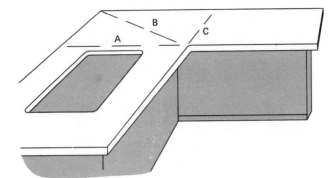

Should you decide to make your own backsplash and countertop, decide which of the three joints you're going to use. The straight joints, *A* and *C*, will be easier to handle than will *B*.

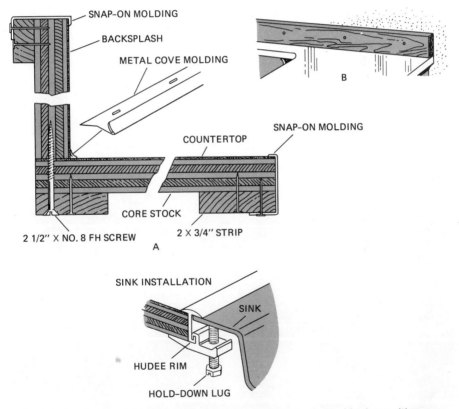

Fabricate and cover backsplash and countertop separately. Assemble them with screws driven up through the countertop as shown in sketch A. Alternatively, you can install the backsplash first, as in sketch B, and butt the countertop to it by reversing the metal cove molding. Either way, be sure to caulk the backsplash-countertop joint. Most laminate dealers have snap-on moldings like those shown. If you cannot find them, self-edge the splash and top. Installation of the sink is carried out with a special fixture called a Hudee Rim, which is a standard plumbing item.

Hydrogen peroxide (in many hair bleaches and dyes)
Argyrol (silver protein; in many eye drops)
Bluing (an additive in some powdered soaps)
Iodine (in any solution)
Potassium permanganate
Gentian violet (any concentration)
Mineral acids: hydrochloric, sulfuric, nitric
Berry juices, such as grape and raspberry
Silver nitrite (any concentration greater than 1%)

To the foregoing list should be added any kind of abrasive scouring material such as steel wool, and cleaners containing abrasive particles, such as Ajax, Dutch Cleanser, etc. In the case of scouring powders, it's not the chemicals used in them in most cases, but the abrasive particles they contain. Continued use will wear away the surface of laminates not specially formulated to resist such wear.

From the foregoing list, you can easily see that you don't have to be operating a chemical laboratory, or even such a commonplace home activity as a photographic darkroom, to encounter chemicals that have a bad effect on laminate surfaces. There are laminates developed for laboratory use which are highly chemical-resistant, but your local supplier isn't likely to have them in stock.

Probably the kindest treatment you can give laminate surfaces, including tileboard fabricated especially for bathroom and kitchen use, is to wash with a mild detergent solution and rinse with warm water, then dry with a soft cloth.

APPENDIX A

CONVERSION FACTORS

WHITE PINE PATTERN LUMBER

Nominal Size	Width Overall	Face	Conversion Factor	Nominal Size	Width Overall	Face	Conversion Factor
SHIPLAP				**TONGUE AND GROOVE**			
1×6	$5^1/_2$	$5^1/_8$	1.17	1×4	$3^3/_8$	$3^1/_8$	1.28
1×8	$7^1/_4$	$6^7/_8$	1.16	1×6	$5^3/_8$	$5^1/_8$	1.17
1×10	$9^1/_4$	$8^7/_8$	1.13	1×8	$7^1/_8$	$6^7/_8$	1.16
1×12	$11^1/_4$	$10^7/_8$	1.10	1×10	$9^1/_8$	$8^7/_8$	1.13
				1×12	$11^1/_8$	$10^7/_8$	1.10
PANELING PATTERNS							
1×6	$5^7/_{16}$	$5^1/_{16}$	1.19	**S4S**			
1×8	$7^1/_8$	$6^3/_4$	1.19	1×4	$3^1/_2$	$3^1/_2$	1.14
1×10	$9^1/_8$		1.14	1×6	$5^1/_2$	$5^1/_2$	1.09
1×12	$11^1/_8$	$10^3/_4$	1.12	1×8	$7^1/_4$	$7^1/_4$	1.10
				1×10	$9^1/_4$	$9^1/_4$	1.08
BEVEL SIDING				1×12	$11^1/_4$	$11^1/_4$	1.07
1×4	$3^1/_2$	$3^1/_2$	1.60				
1×6	$5^1/_2$	$5^1/_2$	1.33				
1×8	$7^1/_4$	$7^1/_4$	1.28				
1×10	$9^1/_4$	$9^1/_4$	1.21				
1×12	$11^1/_4$	$11^1/_4$	1.17				

*Allowance for trim and waste should be added.

REDWOOD PATTERN LUMBER

TONGUE-AND-GROOVE

Sizes	Pattern Numbers	Conversion Factors	Sizes	Pattern Numbers	Conversion Factors
T & G (VIS-S2S)			**La Honda Paneling T & G (S1S)**		
$^3/_4$×6	711	1.15	$^3/_8$×4	S-602	1.19
$^3/_4$×8	712	1.11	$^3/_8$×6	S-603	1.12
			Coverage Per Carton		
			4" width: 67 sq. ft. 6" width: 64 sq. ft.		
Sierra Decking			**Random Plank La Honda Paneling**		
2×6	486	1.17	$^3/_8$×4	S-602	1.19
			$^3/_8$×5	S-605	1.14
Three Way Rustic			*Coverage Per Package*		
$^3/_4$×6	711R	1.15	4" width, 5" width,		
$^3/_4$×8	712R	1.11	8' length: 67 sq. ft. length: 52 sq. ft.		
$^3/_4$×10	713R	1.09	10' length: 84 sq. ft. 10' length: 65 sq. ft.		
			12' length: 100 sq. ft. 12' length: 78 sq. ft.		

(continued)

APPENDIX A

CONVERSION FACTORS *(continued)*

	BEVEL SIDING PATTERNS			SHIPLAP PATTERNS	
Sizes	*Pattern Numbers*	*Conversion Factors*	*sizes*	*Pattern Numbers*	*Conversion Factors*

Plain Bevel Siding (S1S2E)   V RUSTIC (S2S)

	BEVEL SIDING PATTERNS			SHIPLAP PATTERNS	
$1/_2 \times 6$	322	1.34	$3/_4 \times 6$	793	1.20
$1/_2 \times 8$	323	1.24	$3/_4 \times 8$	794	1.15
$1/_2 \times 10$	324	1.18			

Bevel Siding (S1S2E)   Channel Rustic (S2S)

$3/_4 \times 6$	329	1.34	$3/_4 \times 6$	774	1.20
$3/_4 \times 8$	330	1.24	$3/_4 \times 8$	775	1.15
$3/_4 \times 10$	331	1.18	$3/_4 \times 10$	776	1.12

Rabbeted Siding (S1S2E)

$1/_2 \times 6$, $^{15}/_{32}''$	L362	1.20
$1/_2 \times 8$, $^{15}/_{32}''$	L363	1.15
$3/_4 \times 8$, $^{11}/_{16}''$	372	1.15
$3/_4 \times 10$,	373	1.11
$3/_4 \times 10$, $^{11}/_{16}''$	392	1.13
$3/_4 \times 10$ Rough $3/_4''$	393	1.10

APPENDIX B

AMERICAN STANDARD SOFTWOOD LUMBER DIMENSIONS

		THICKNESS		FACE WIDTH		
	Nominal	Dry	Green	Nominal	Dry	Green
	inches	*inches*	*inches*	*inches*	*inches*	*inches*
BOARDS AND STRIPS	1	$3/4$	$25/32$	2	$1^1/_2$	$1^9/_{16}$
	$1^1/_4$	1	$1^1/_{32}$	3	$2^1/_2$	$2^9/_{16}$
	$1^1/_2$	$1^1/_4$	$1^9/_{32}$	4	$3^1/_2$	$3^9/_{16}$
				5	$4^1/_2$	$4^5/_8$
				6	$5^1/_2$	$5^5/_8$
				7	$6^1/_2$	$6^5/_8$
				8	$7^1/_4$	$7^1/_2$
				9	$8^1/_4$	$8^1/_2$
				10	$9^1/_4$	$9^1/_2$
				11	$10^1/_4$	$10^1/_2$
				12	$11^1/_4$	$11^1/_2$
				14	$13^1/_4$	$13^1/_2$
				16	$15^1/_4$	$15^1/_2$
DIMENSION LUMBER	2	$1^1/_2$	$1^9/_{16}$	2	$1^1/_2$	$1^9/_{16}$
	$2^1/_2$	2	$2^1/_{16}$	3	$2^1/_2$	$2^9/_{16}$
	3	$2^1/_2$	$2^9/_{16}$	4	$3^1/_2$	$3^9/_{16}$
	$3^1/_2$	3	$3^1/_{16}$	5	$4^1/_2$	$4^5/_8$
	4	$3^1/_2$	$3^9/_{16}$	6	$5^1/_2$	$5^5/_8$
	$4^1/_2$	4	$4^1/_{16}$	8	$7^1/_4$	$7^1/_2$
				10	$9^1/_4$	$9^1/_2$
				12	$11^1/_4$	$11^1/_2$
				14	$13^1/_4$	$13^1/_2$
				16	$15^1/_4$	$15^1/_2$
TIMBERS	5 and greater	$1/_2$" less than nominal		5 and greater	$1/_2$" less than nominal	

Example of use: You order 2×4 dimension lumber (nominal dimension). You get the "dry" or "green" thickness and face widths shown in those columns. A 2×4 from a lumberyard is $1^1/_2×3^1/_2$ inches.

APPENDIX C

LUMBER AND BUILDING-INDUSTRY TRADE ASSOCIATIONS

Here are the trade associations which you can write for detailed information regarding lumber, plywood, and panelboards. Most of the associations have how-to-do-it folders or booklets that may help you with specific jobs you have in mind.

HARDBOARDS:

American Board Products
Association
205 West Touhy Avenue
Park Ridge, IL 69968

PLYWOOD:

American Plywood Association
1119 A Street
Tacoma, WA 98401

Hardwood Plywood Manufacturers
Association
P. O. Box 6246
Arlington, VA 22206

TREATED WOOD:

American Wood Preservers Institute
1651 Old Meadow Road
McLean, VA 22101

HARDWOODS:

Appalachian Hardwood
Manufacturers, Inc.
N. C. Nat'l Bank Bldg.
High Point, NC 27260

REDWOOD:

California Redwood Association
617 Montgomery Street
San Francisco, CA 94111

**STANDARDS AND
SPECIFICATIONS:**

Construction Specifications Institute
1150 17th Street NW
Washington, DC 20036

EXOTIC WOODS:

Imported Hardwood Products
Association
P. O. Box 1308
Alexandria, VA 22313

BUILDING DATA:

National Association of Home
Builders
15 & M Streets
Washington, DC 20005

LAMINATES:

National Association of Plastic
Fabricators
1701 N Street
Washington, DC 20036

RETAIL DISTRIBUTION:

National Lumber and Building
Material Dealers
1990 M Street NW
Washington, DC 20036

WOOD FINISHING:

National Paint and Coatings
Association
1500 Rhode Island Avenue NW
Washington, DC 20005

GYPSUM BOARD:

United States Gypsum Company
101 S. Wacker Drive
Chicago, IL 60606

PARTICLEBOARD:

National Particleboard Association
2306 Perkins Place
Silver Spring, MD 20910

NORTHERN HARDWOODS:

Northeastern Lumber
Manufacturers Association
5 Fundy Road
Falmouth, ME 04105

SOUTHERN SOFTWOODS:

Southern Forest Products
Association
P. O. Box 52468
New Orleans, LA 70152

SOUTHERN HARDWOODS:

Southern Hardwood Lumber
Manufacturers Association
805 Sterick Bldg.
Memphis, TN 38103

WESTERN SOFTWOODS:

Western Wood Products Association
1500 Yeon Bldg.
Portland, OR 97205

Index